THE VERTEBRATES OF ARIZONA

THE VERTEBRATES OF ARIZONA

landscapes
and habitats
fishes
amphibians
and reptiles
birds
mammals

CHARLES H. LOWE
editor

THE UNIVERSITY OF ARIZONA PRESS
Tucson, Arizona

First Printing 1964
Second Printing 1964
Third Printing 1967
Fourth Printing 1972

THE UNIVERSITY OF ARIZONA PRESS

Manufactured in the U.S.A.

I. S. B. N.-0-8165-0348-6
L. C. No. 63-11981

PREFACE

A symposium on "The Vertebrates of Arizona" was held at the University of Arizona, Tucson, on April 9, 1960, at the annual meeting of the Arizona Academy of Science. As Chairman of the Biology Section of the Academy during that year, I arranged the symposium as a departure point for the publication of needed check lists of the Recent vertebrates of Arizona. Accordingly, check list summaries and analyses for the vertebrates presently occurring within the state were presented by E. Lendell Cockrum (mammals), Gale Monson and Allan R. Phillips (birds), Charles H. Lowe (amphibians and reptiles), and Robert R. Miller and Charles H. Lowe (fishes). Howard K. Gloyd, pioneer investigator of Arizona's vertebrate fauna, opened the symposium with a review of some of the historical highlights in the biological exploration of the state.

Check lists have continued to be requested by a growing number of people with diverse interests and needs. Since the summer of 1957, when Arizona's first National Science Foundation science institute for high school teachers was held at The University of Arizona, high school teachers of Arizona and other western states have constituted one of the major groups that have sought an account of Arizona's present vertebrate fauna. Thus the needs of teachers in general have been kept in mind throughout this work.

The check lists are accounts which constitute an up-to-date concise inventory of the *species* in the Recent vertebrate fauna of the state. The *subspecies* are not included, and only an occasional pertinent reference to some subspecies is made. Comprehensive treatments of the taxonomy and/or ecology of each of the vertebrate groups as they occur in Arizona and inclusive of subspecies *et al.*, have been in preparation for a number of years. One of these works was recently published by Cockrum (1960; The Recent Mammals of Arizona).

In the symposium an introduction to biogeographical and ecological aspects of the fauna was given in the form of an ecological check list, *i.e.*, a brief survey of the major types of land environments available to and occupied by vertebrate animals in Arizona, from desert and grassland through woodland and forest to alpine tundra. This is included in the present work (Part I) as are some maps, one of which also locates major waters supporting Arizona's native and introduced fish fauna.

Some of the photographs in Part I are of Arizona landscapes and habitats as they were observed roughly a half-century ago. The prints were made from the original glass plate negatives which are deposited in the McDougal-Shreve Ecology Collection housed in the Department of Zoology, Biological Sciences Building, University of Arizona.

I think it appropriate indeed that some of these very excellent photographs, representing Arizona during many years past as well as today, are used in the present volume; however, it is certainly true that some significant changes have taken place in the vegetation and habitats of Arizona during the past half-century, and, therefore, I should briefly summarize them here with regard to the MacDougal-Shreve photographs which I have used.

1. A *Sonoran Desert* landscape photographed in 1915 is shown in Figure 6. Reexamination of this locality today reveals that few and relatively minor vegetation changes have taken place on this particular desert site during the past half-century.

2. *Desert-grassland* landscapes are shown in Figures 27 (1912) and 33 (1911). Today the desert-grassland exhibits marked shrub and tree "invasion" by dry-tropic species at the expense of the perennial native grasses; this has been a relatively long term disturbance-response triggered near the turn of the century.

3. *Evergreen woodlands* are shown in Figures 33, 35, and 36. Many of the oaks occurring along the open lower edges of these woodlands (transitional to grassland) in southern Arizona have recently died as a result of critically lowered soil moisture content, just as many similar live-oaks have died elsewhere since the early 1940's — across the Southwest from southern California eastward into Texas. However, the trees *within* the woodland remain relatively little affected (*e.g.*, Fig. 33, 1911) and the oak woodland itself is thus little changed by these more recent environmental changes initiated just before the turn of the half-century.

4. Some of Forrest Shreve's early photographs of the *coniferous forest* on the Santa Catalina Mountains, near Tucson, are shown in Figures 44-47. Uncut and relatively ungrazed coniferous forest climax (as in Fig. 46, 1908) that is highly elevated on rugged mountainous terrain reveals little significant change in its vegetation over the past half-century in Arizona.

Each of the five parts in the present work has been constructed with a view towards its additional availability as a separate. It is a pleasure to acknowledge the assistance of The University of Arizona Press in this work, and in particular the help of Dr. H. K. Gloyd of the University of Arizona Department of Zoology.

C. H. Lowe
Editor

CONTENTS

Part 1

Part 2

Part 3

Part 4

Part 5

FIGURES

(Photographs are by the authors unless otherwise credited)

TABLES

PART I

ARIZONA LANDSCAPES AND HABITATS

Charles H. Lowe

Introduction

The variety and beauty of the landscapes of Arizona have been remarked upon by naturalists who have long-traveled to this remarkable state from many parts of the World. This introduction to the Arizona vertebrate animal check lists provides an illustrated account of these natural settings for the homes of animals and plants in the state, and a discussion of the systems employed by biologists to classify them. Some references are given to aid those interested in further exploration of the subject.

Arizona is not only a desert. It is one of the most topographically varied and scenic areas in North America and contains extensive forests occurring up to timberline elevations (11,000-11,400 feet) plus a bit of alpine tundra above timberline on the summit of San Francisco Mountain (to 12,670 feet). This great environmental diversity is reflected in Arizona's diverse fauna in which there is a total of 64 species of fishes, 22 species of amphibians, 94 species of reptiles, 434 species of birds, and 137 species of mammals, a combined total of 751 species which does not include several others listed as "hypothetical" for Arizona.[1]

Habitats

The vertebrates occupy the two major kinds of habitats, those on land and those in water. The general term *habitat* refers to the natural environs of a species of plant or animal — the place where it lives and makes its living. Habitats may be small (microhabitats) or large (macrohabitats). The latter may be large community habitats where many species of plants and animals live — the habitats of biotic communities — and each species occupies its own ecologic niche within the community.

Habitats themselves, just as plants and animals, reveal certain similarities and differences. Because of this, a check list of habitats can be arranged (Table 1) that is similar to a check list of animals, say of birds. Thus, in addition to the *taxonomic classification* (*systematic hierarchy*) for animals and plants (*e.g.*, species, genus, family, order, etc.) which is based on phylogeny and heredity,[2] there is an *ecologic classification* (*e.g.*,

---

[1] Numbers are based on the check lists of this series (1964).

[2] See Mayr, Linsley, and Usinger (1953) for principles and procedures in classification. Recent symposia entitled "The Species Problem" (Mayr, 1957), "Species: Modern Concepts" (Lewis, Maslin, Durrant, Phillips, Lowe, 1959), and "Vertebrate Speciation" (Blair, 1961) contain papers by western naturalists which treat species problems in Arizona and other southwestern states, in addition to the general problem of the mechanisms in the formation of species.

association, association-type, formation, formation-class, etc.) based primarily on life-form and environment, for example as follows:[3]

Woodland Formation-class
  Evergreen Woodland Formation
    Oak Woodland (*Quercus*) Association-type
      Emory oak-Arizona oak Association

While ecologic classification is natural, it lacks the unifying theme of phylogenetic relationship that is found in taxonomic classification. Of course, natural landscapes do evolve and have evolutionary relationships of their own. The story of the evolution of vegetation, *e.g.*, of our forests, grasslands, deserts, etc., is one of the most fascinating stories in ecology and evolution (see Chaney, Dorf, and Axelrod, 1944; Axelrod, 1950, 1958; Deevey, 1949; Dorf, 1960; Martin, Schoenwetter and Arms, 1961). Moreover, it is obvious that *natural selection* (Darwin, 1859) is an ecological concept.

Plants have left relatively abundant fossil remains (as compared to animals) and these are frequently in the form of leaves (megafossils) and pollen (microfossils). This has proved important to modern paleoecologic investigation of the evolution of biotic communities. For example, investigation by the floristic method which aids reconstruction of past environments by comparison of a fossil flora with the living vegetation which resembles it most closely (see references cited above for Chaney, *et al.*, and Axelrod). Norris (1958), Gray (1961), and Martin and Gray (1962) have recently analyzed Pleistocene events in the arid Southwest and events in the West during late Tertiary. Darrow (1961) has recently discussed the origin and development of the vegetational communities of the Southwest, and Turner (1959) also reviewed Axelrod's work with special reference to this area.

Why is the ecologic classification of terrestrial biotic communities based essentially on plants? The answer is not hard to appreciate. Plants, in their varied forms and quantitative relationships, constitute one of the essential features of landscape, and hence of landscape classification. The plants are always out there "taking it" — for 24 hours a day for all 365 days of the year. They continuously respond by species, numbers, sizes, sites, etc. to the factors controlling the biotic (plant-animal) community. Thus they most faithfully express the *effective environment* to the observer. Animals are more or less secretive by day or by night, and often indeed for one or more entire seasons. There is no question that the vegetation

---

[3] This is a conventional form of ecologic classification. See Fosberg (1961) for a different scheme.

The smallest possible unit in such an ecological classification, the ecological niche, is not included here. Each species within the "association" occupies its own peculiar ecologic niche ("address" + "occupation") in the biotic community.

offers for the present the most satisfactory basis for the recognition and classification of the major terrestrial communities.

It is the *native perennial plants* that are the ever-present, sensitive, readily observed, measureable and mappable indicators of the environmental controls, which are climate, soil, topography, and biotic factors including man, his domestics, and man-made fire.[4] The perennials constitute the important fraction of the plant-animal community that we perceive as vegetation and classify in any meaningful hierarchy of natural communities. In short, as vegetation, native perennial plants are the classifiable and mappable indicators of the environments also inhabited by the animals; hence by the entire biotic community.

Thus, by and large, it is the plants rather than the animals that tell us more precisely what the overall environmental conditions were at a given place and time in the geological yesterday, as well as what the effective environments are in a given area today.

The following illustrated references are available and particularly useful for aid in recognizing native plants in Arizona: Benson (1940), Benson and Darrow (1944), Preston (1947), Little (1950), Gould (1951), Humphrey, Brown, and Everson (1956). Kearney and Peebles' (1942, 1951, 1960) flora for Arizona is a more extensive work; the more comprehensive flora for California (Munz and Keck, 1959), which includes a large number of species common to both states, is more recent and considerably better from an evolutionary point of view (*Larrea divaricata* for creosotebush, *Cereus giganteus* for sahuaro, etc).[5] Little's (1953) checklist is the most recent available for trees. Standard and illustrated works on North American trees are Martinez (1945, 1948) for pines, Sargent's (1926) manual, Sudworth's (1915 to 1934) series of papers on species of the Rocky Mountain region, and Munn's (1938) maps. The popular series on flowers published by the Southwest Monuments Association are useful (Dodge, 1958; Arnberger, 1957; Patraw, 1953).

Pearson's (1931 and elsewhere) and Shreve's (1942a and elsewhere) accounts of the vegetation of Arizona are excellent, and Nichol's (1937, 1952) vegetation map is still serviceable after 25 years (although in need of revision) as are the map and the classifications on which it was based (Shreve, 1917; Livingston and Shreve, 1921; Shantz and Zon, 1924; Shantz, 1936; and others). Humphrey's (1950, 1953, 1955, 1960) illustrated papers with maps of Arizona grassland vegetation are available for many counties; Darrow (1944), for Cochise County. Other maps

---

[4] For discussions of fire and Arizona vegetation, see Leopold (1924), Pearson (1931), Humphrey (1949, 1951, 1958, 1962), Reynolds and Bohning (1956), Marshall (1957, 1962), and Cooper (1960, 1961b).

[5] The scientific names for plants used here may be found (with few exceptions) in the following works which, of course, are not always in agreement: Little (1953), Benson (1940), Benson and Darrow (1944, 1954), Munz and Keck (1959), Kearney and Peebles (1951, 1960).

covering parts of Arizona are provided by Bryan, 1925; Baker, 1945; U. S. Forest Service, 1949; Marshall, 1957; Martin, *et al.*, 1961; and others (see Melton, 1959, for a map series illustrating a geomorphic history of southeastern Arizona). Topographic, climatic, and other maps are in the recent and available volume, *Arizona, Its People and Resources* (Cross, 1960). Fenneman (1931) is a standard general reference for physiography in the West. Historic accounts, usually illustrated, of the more recent changes in Arizona's landscapes and natural vegetation cover are given by several investigators.[6] There are numerous references (in addition to those noted above) which are primarily descriptive ecological studies on the vegetation and/or flora of Arizona.[7]

Rydberg (1913-1917) published a series of seven notable papers in the Bulletin of Torrey Botanical Club, entitled "Phytogeographical Notes on the Rocky Mountain Region." The extensive publications of the Rocky Mountain Forest and Range Experiment Station, Ft. Collins, Colorado (formerly at Tucson, Arizona), should be consulted in addition to those station papers referred to here.

Succession

The term "forest type" is generally used by foresters for a unit characterized by uniformity in composition of tree species; and "vegetation type" for any form of vegetation whether tree-form or not. "Forest type" (and "vegetation type") is equivalent to the ecological term "association" and to "consociation" where only one species is dominant. Figure 46 is of

---

[6] For example, Bryan (1928), Shreve (1929), Pearson (1931), Shreve and Hinkley (1937), Parker (1945), Brown (1950), Glendening (1952), Parker and Martin (1952), Humphrey (1953b, 1958), Schroeder (1953), Arnold and Schroeder (1955), Schulman (1956), Humphrey and Mehroff (1958), Hastings (1959), Lowe (1959a), Murray (1959), Cooper (1961), Yang (1961), Marshall (1962), and others. See Campbell and Bomberger (1934), Norris (1950), Gardner (1951), Branscomb (1958), Yang (1961), and others, for adjacent New Mexico.

[7] "Lowlands" refers here to desert and grassland habitats; "Highlands" refers to chaparral, woodland, forest, and alpine tundra. *Lowlands:* Loew (1875), Lloyd (1907), Thornber (1910), Spalding (1909, 1910), Harshberger (1911), Blumer (1912), Shreve (1915, 1925, 1942b, 1951), Aldous and Shantz (1924), Hanson (1924), Shantz and Zon (1924), Shantz and Piemeisel (1925), Sturdevant (1927), McKee (1934), Gloyd (1932, 1937), Woodbury and Russell (1945), Munz and Keck (1949), Spangle (1949), Leopold (1950), Merkle (1952), Sutton (1952), Deaver and Haskell (1955), Wallmo (1955), Yang and Lowe (1956), Yang (1957), Haskell (1958), Keppel, *et al.* (1958-60), Lowe (1959a, 1961). *Highlands:* Loew (1875), Hoffman (1877), Britton (1889), Rusby (1889), Leiberg, Rixon and Dowell (1904), Blumer (1909, 1910, 1911), Harshberger (1911), Woolsey (1911), Read (1915), Shreve (1915, 1919), Eastwood (1919), Pearson (1920, 1931, 1933, 1942, 1950), Hanson (1924), McHenry (1932, 1933, 1935), Croft (1933), Dodge (1936), Gloyd (1932, 1937), Howell (1941), Little (1941), Rasmussen (1941), Martin and Fletcher (1943), Baker (1945), Woodbury and Russell (1945), Arnberger (1947), Woodbury (1947), Peattie (1953), Dickerman (1954), Wallmo (1955), Marshall (1956), Cooper (1961a), Lowe (1961), Jameson, Williams and Wilton (1962).

Jaeger's book on The North American Deserts (Stanford University Press, 1957) is based importantly, and without due credit, on the work of Forrest Shreve.

an *association* that is a common forest type, the fir type (or Douglas fir type). Figure 48 is of a *consociation* that is a common forest type, the aspen type (or quaking aspen type).

Foresters also use the terms "temporary" and "permanent" for *successional* and *climax* stages, respectively. Thus the aspen type (Fig. 48) is a temporary forest type and the fir type (Fig. 46) is a permanent type. Aspen is temporary because it will be replaced in time by conifers, to form a permanent fir type, spruce type, or pine type, as the case may be. The succession takes place after the "permanent" climax trees have been removed, as by fire, logging, etc.; grasses ordinarily establish first, followed by the aspens and eventually by the conifers. Succession is the rule in the forest, woodland, and grassland, but not in desert. In the desert, succession is the rare exception (see Shreve, 1925; Lowe, 1959a).

The Climax Pattern

The climax is the particular pattern of the recognizably mature, characteristic vegetation and associated animals of the biotic community; that is, the natural climax pattern, or climax community pattern. The climax pattern of a geographic area is comprised of one, two, or more climax biotic associations, all of which are characterized by shared distinctiveness in *life-form* (or of strikingly different life-form) of the important climax species. Each of these biotic associations (when there is more than one) varies obviously from the other(s) in the *species composition* of its own particular climax community.

Thus there may be (1) only one climax biotic association in the area that is present on all sites, as is so often the case within areas of extensive forest (*e.g.*, in western spruce-alpine fir forest, eastern beech-maple forest); or, (2) the climax pattern may have a larger inherent genetic variation and be comprised of two or more climax biotic associations which may vary greatly in species composition (and possibly life-form) from one physical habitat site to another adjacent one within a given geographic area under the same climate; this is usually the situation in the desert (*e.g.*, Sonoran Desert, Fig. 7 and 8; see Yang and Lowe, 1956).

The climax pattern of a forest on a mountain is likely to be in the form of a *continuum* on the moisture-temperature gradient that is present; that is, a continuum of gradually changing genetic composition and life-form complexity from one extreme of the environmental gradient to the other. A climax pattern in the desert is just as likely to be one of a *mosaic* of climax species and life-forms (biotic association) as it is to be a continuum; that is, a mosaic of abruptly changing genetic composition and life-form complexity under the same climate. Hence often, as in the desert, there is abrupt repetition of climax biotic associations in the emerging form of a huge and irregular environmental chessboard on which the plants and animals are the pawns of the paired controls of topography and soil under the same climate.

This individualistic nature of the community and climax pattern was first reported very astutely and clearly by Forrest Shreve (1914, 1915) nearly a half-century ago (see also Shreve 1917, 1919, 1925, 1951). It was also ably championed by Gleason (1917, 1926, and elsewhere), Law (1929), and others in this country and abroad. But not until relatively recently has it been comprehended (or admitted) widely among American ecologists (Cain, 1939; Muller, 1939, 1940; Mason, 1947; Egler, 1951; Curtis and McIntosh, 1951; Whittaker, 1951, 1957; Brown and Curtis, 1952; Lowe, 1959a). For the somewhat older, organismic idea in American ecology, see Weaver and Clements (1938) or Allee, *et al.* (1949); also Kendeigh (1961).

The Climate

An understanding of the biotic communities, and the life-zones in Arizona which they comprise, is facilitated by an understanding of the physical factors which ultimately control them, particularly those of climate. A brief consideration of climate and other controlling environmental factors is presented at various places below (see Sequence of Biotic Communities and Zones). Kincer (1941), Smith (1956), Keppel, *et al.* (1958-60), Sellers (1960a, 1960b), and Green (1961) provide detailed climatological data and information for Arizona. *The Climate of Arizona* by Sellers (1960b) is a particularly informative paper on the weather elements and pertinent topographic features of the state (Fig. 1).

All of Arizona falls under the Southwestern or Arizona climatic pattern (Kincer, 1922), which is a bi-seasonal regime characterized by winter precipitation, spring drought, summer precipitation, and fall drought (Reed, 1933, 1939; Alexander, 1935; Holzman, 1937; Turnage and Mallory, 1941; Dorroh, 1946; Ives, 1949; Jurwitz, 1953; Sellers, 1960a). For most plants and animals the fore-summer drought which is associated with higher temperatures (May-June) is the more severe of the two drought periods.

The moisture for the state's bi-seasonal regime comes from northerly directions in the winter and from southerly directions in the summer. The summer precipitation comes from storms which are primarily convectional in nature, often intense, and characteristically local rather than widespread, with most storms having a diameter of less than three miles. This results from moist tropical air which moves into the state from the southeast (Gulf of Mexico, Atlantic Ocean) and the southwest (Gulf of California, Pacific Ocean) and then passes over strongly heated and mountainous terrain which causes it to rise rapidly, cool, and condense.

Summer rainfall in Arizona is from mid-May to mid-October, although the two driest months (May and June) are usually rainless. The summer monsoon begins dramatically in early July (or, less frequently, in late June), suddenly breaking the fore-summer drought and Arizona's

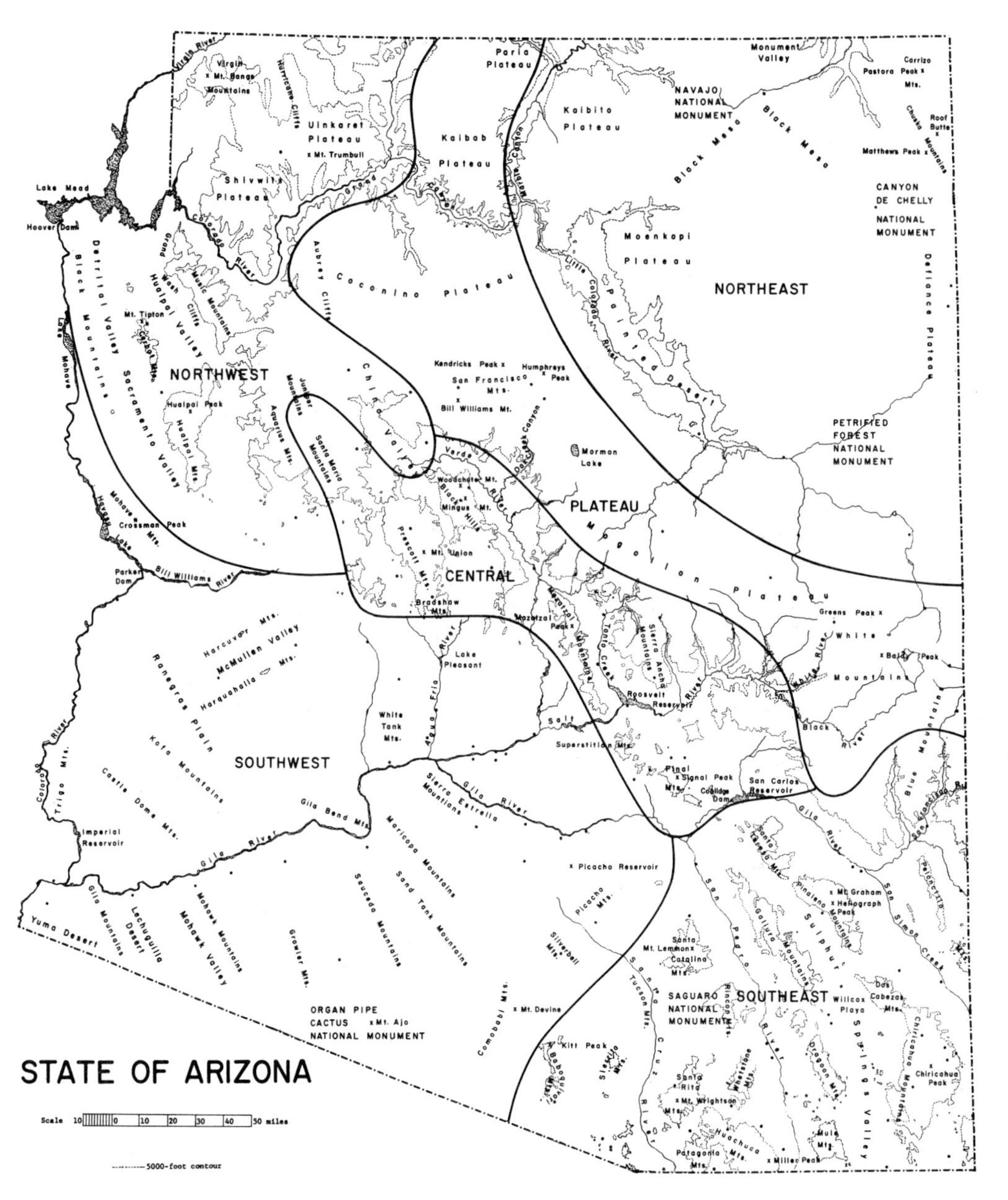

Fig. 1. Topographical features of Arizona and section boundaries used for discussion of the climate of Arizona by Sellers (1960b).

hottest weather of June-July. The major source of this precipitation throughout the summer is the Tropical Atlantic (Gulf) air mass, but in late summer (late August, September, early October) surges of moisture also may move into the state from the Tropical Pacific and contribute sizeable storms. These may be the fringes of Mexican west coast hurricanes.

The winter precipitation is associated with the westerlies which normally bring Polar Pacific air moisture onto the continent in Washington and Oregon, and occasionally do so as far south as central California. Accordingly, the convergent winter storms from the northwest (NW, N, W) may or may not pass over Arizona. When they do, surface thermal heating is obviously much less pronounced than in the case of summer conditions, upslope air movement is relatively slow, cloudiness is widespread, and the precipitation tends to be more gentle and to cover more area per storm.

In spite of the seasonal differences in storm types which are predominantly general in winter and local and erratic in summer, the winter precipitation in Arizona is decidedly more variable from year to year; it is more variable both in amount and in time of occurrence than that of the summer with its remarkably sharp onset in early July (McDonald, 1956).

In Arizona and neighboring southwestern areas, the total precipitation increases on mountain gradients at a rate of approximately 5 (4-5) inches per 1,000 feet increase in elevation (Shreve, 1915; Pearson, 1920; Sykes, 1931, Hart, 1937; Schwalen, 1942; Lull and Ellison, 1950). At the higher and usually more humid elevations, however, yearly variation is less than at the lower and ordinarily more arid levels; this relationship is also seen when comparing the eastern (more humid) and western (more arid) United States.

ECOLOGIC CHECK LIST

An ecologic check list for Arizona is given in Table 1. It is illustrated in Figures 2-53, and is annotated in the captions of the figures. Six world *ecological formation-classes*, and ten major subcontinental ecological *formations* of North America, are represented in the landscapes of Arizona.[8]

World Formation-classes

Desertscrub, grassland, chaparral, woodland, forest, and tundra landscapes are found on various continents throughout the world. These are the major types of ecological (=plant-animal, or biotic) formations, *i.e.*, the formation-classes are the principal *biotic communities* of the world. As usual in ecology, they are classified primarily on the basis of vegeta-

[8] For the world phytogeographic picture see Cain (1944), Good (1947), and Dansereau (1957). For the zoogeographical realms of the world (Sclater, 1858; Wallace, 1876) see Darlington (1957) and, for a brief outline, Storer and Usinger (1957).

tion (and climate) rather than animals.[9] They are occasionally called "biomes," or "biome-types."

Desertscrub

Arid, hot (or hot and cold; or cool) environments with irregular winter rainfall, summer rainfall, or bi-seasonal rainfall, which vary from (1) open, often-thorny (spinose) microphyllous *short-tree and shrub* (and other scrub) vegetation; (2) open, well-spaced microphyllous *shrub* (and other, often-thorny scrub) vegetation predominant or exclusive; to (3) *none*. Plant life-form is highly varied with leafless, drought deciduous, and evergreen species, including trees and shrubs, herbs and grasses, yuccas and agaves, cacti and ocotillo, and other, occasionally bizarre, forms.

Grassland

Semi-arid to semi-humid, warm to cold environments: (1) semi-arid *steppe, plains,* or *desert-grassland* with short grasses predominant, and occasionally scattered scrub (shrubs, yuccas, agaves, cacti, etc.); (2) semi-arid to semi-humid *savana grassland* with scattered trees and shrubs; (3) semi-humid *prairie grassland* of tall and/or mid grasses; (4) semi-humid and low temperature *mountain grassland* (mountain meadow) with short grasses and sedges.

Chaparral

Semi-arid (usually), warm to cool environments with dense, and usually closed, short-statured, mostly shrubby evergreen vegetation dominated by usually sclerophyllous, small-leaved to broad-leaved (1) shrubs, or (2) shrubs and dwarf trees (commonly scrub oaks); one species commonly dominates.

Woodland

Semi-arid to semi-humid, warm to cold environments with a more or less open canopy of (1) *evergreen* trees which are primarily species of oak, juniper, pinyon pines, and other similar life-forms, with an undergrowth of grasses and (a) woodland evergreen (often deciduous) shrubs and dry-tropic (usually deciduous) scrub, and/or (b) desert scrub (evergreen, deciduous, or leafless shrubs, cacti, yuccas, etc.); or (2) *deciduous broadleaf riparian* trees such as species of cottonwood, willow, ash, walnut, sycamore, alder, chokecherry, mulberry.

Forest

Semi-humid (occasionally semi-arid) to wet, and cold to warm environments with a more or less closed stand of trees forming a more or less continuous canopy: (1) *tropical rain forest* of tall stature, with variously shaped luxuriant foliage, in more or less continuously warm and

---

[9] In addition, I would follow Dansereau (1957) and consider the formation-classes as major subdivisions (or equivalents in part) of earth biochores; but of these three — rather than four — as follows: Forest biochore, Savana biochore, and Desert biochore.

Table 1. Check List of Biotic Communities in Arizona and Their Major Subdivisions

Desert Formation-class
- 1. *Southwestern Desertscrub* Formation*
  - Creosotebush associations
  - Tarbush "
  - Whitethorn "
  - Sandpaperbush "
  - Joshuatree "
  - Blackbrush "
  - Saltbush "
  - Paloverde "
  - Mesquite "
- 2. *Great Basin Desertscrub* Formation†
  - Sagebrush associations
  - Blackbrush "
  - Shadscale "

Grassland Formation-class
- 3. *Desert-Grassland* Formation‡
  - Desert-Grass associations
- 4. *Plains Grassland* Formation
  - Shortgrass Plains associations
- 5. *Mountain Grassland* Formation
  - Mountain Grass associations

Chaparral Formation-class
- 6. *Chaparral* Formation
  - Interior Chaparral associations

Woodland Formation-class
- 7. *Evergreen Woodland* Formation
  - Oak Woodland associations
  - Oak-pine Woodland associations
  - Juniper-pinyon Woodland associations
- 8. *Decidious Woodland* Formation
  - Riparian Woodland associations

Forest Formation-class
- 9. *Coniferous Forest* Formation
  - Ponderosa Pine Forest associations
  - Douglas Fir Forest "
  - Limber Pine Forest "
  - Spruce-alpine fir Forest "
  - Aspen associes "

Tundra Formation-class
- 10. *Alpine Tundra* Formation
  - Alpine Tundra associations

wet environments; (2) *tropical deciduous forest* of short to medium stature, with broad and small leaves that fall during the *dry* season; (3) *temperate deciduous forest* with broad leaves that fall during the *cold* season; (4) *broad-leaved evergreen forest* with variously textured broad leaves, irregular leaf-fall and generally mild cold season; (5) *needle-leaved evergreen (coniferous) forest* dominated by pines, firs, spruces, hemlocks, etc., in ordinarily semi-humid or humid, temperate to cold environments.[10]

Tundra

Cold, treeless, high latitude and high elevation environments: (1) *arctic tundra* vegetation (primarily sedges, lichens, mosses, grasses, herbs, and low shrubs) and associated animals occur characteristically over extensive plains-like landscapes with an underlying soil permafrost, and under an arid to semi-arid climate, with low precipitation, relatively high atmospheric pressure and oxygen concentration; (2) *alpine tundra* vegetation (primarily sedges, grasses, and herbaceous forms) and associated animals occur above timberline on the tops of higher mountains (at approximately 11,000-11,500 feet elevation in Arizona and New Mexico), on ordinarily shallow, mostly rocky, often unstable soils (without permafrost) under relatively humid climatic conditions with high precipitation and relatively low atmospheric pressure and oxygen concentration.

Rocky Mountain Biotic Communities

The primary characteristic biotic communities in the southern Rocky Mountains (or Southern Rocky Mountain Province) are given in Table 2. The southern Rockies are here considered as extending from their southernmost outliers, for example in southern New Mexico (Sacramento Mountains), southern Arizona (Santa Catalina-Rincons), and southern Nevada (Charleston Mountains), northward to the latitude of mid-Wyoming. All of the southernmost Rockies lie north of the Sierra Madre Occidental, and within the United States. Some of these communities (Table 2) also occur farther northward in the Rockies and some of them southward into Mexico. Table 2 for the southern Rockies may be compared with Table 1

[10] The classification of the world's forests is particularly complicated by their tremendous structural diversity. No brief ecologic classification, of course, is wholly representative of any formation-class.

Footnotes here refer to Table 1, page 12

*Mohave Desert, Sonoran Desert, and Chihuahuan Desert (Shreve, 1942a, 1942b, 1951). These are Scrub associations; those dominated by creosotebush, for example, may be termed the Creosotebush Scrub associations, etc.

†Great Basin Desert (Shreve, 1942b).

‡The mythical Desert-Grassland is listed as a formation merely to follow convention. It is a transitional region (ecotone) with a transitional climate between grassland and desert (Shreve, 1942a, b, c); it is, incidentally, misunderstood by many American ecologists and often misinterpreted.

for Arizona and with Daubenmire's earlier concept of the ecologic formations and "zones" involved (1943b).

A somewhat less than primary but nevertheless characteristic forest type in the Rocky Mountains is also listed for completeness, the limber pine forest (Pearson, 1931). Limber pine and bristlecone pine form conspicuous high elevation pine communities, either as (1) associations with both species present or (2) consociations dominated by one or the other species, as discussed below under spruce-fir forest (Hudsonian Life-zone).

Table 2. The Primary Characteristic Biotic Communities in the Southern Rocky Mountains

See Table 1	Daubenmire, 1943b
Alpine Tundra	Tundra Formation
	Alpine tundra zone
Coniferous Forest	Needle-Leaved Forest Formation
Spruce-alpine fir Forest	Englemann spruce-subalpine fir zone
Limber pine Forest	Douglas fir zone
Douglas fir Forest	Ponderosa pine zone
Ponderosa pine Forest	Juniper-pinyon zone
	Oak-mountain mahogany zone
Evergreen Woodland	
Juniper-pinyon Woodland	
Oak Woodland	
Oak-pine Woodland	
Chaparral	
Interior Chaparral	
Grassland	Grassland and Desert Formations
Mountain Grassland	
Plains Grassland	
Desert-Grassland	
Desert	
Great Basin Desertscrub	
Southwestern Desertscrub	

Both pines reach their southernmost limits in Arizona; limber pine reaches its northerly limit in Alberta, bristlecone pine in Colorado. These communities are characteristic of wind-swept ridges and cold canyon heads at high elevations, and often cover entire slopes at elevations of 10,000 feet and above. Collectively they may be termed limber pine forest or limber pine-bristlecone pine forest (or, Southwestern high pine forest).

Nichol (1937) considered all of Arizona's coniferous forests as a single "Douglas fir — ponderosa pine type." While this obviously does not represent the actual vegetation of Arizona, it was adequate for a map of the vegetation of the entire United States by Schantz and Zon (1924; and Schantz, 1936), who were followed by Nichol without further detail for Arizona.

ARIZONA LIFE-ZONES

The vertically arranged zones of plant and animal life which occur in the region of the San Francisco Peaks in central Arizona were studied by C. Hart Merriam and his associates in 1889. These particular vertical zones of life, ranging between 3,000 and nearly 13,000 feet, were mapped (on the basis of vegetation) by Merriam (1890). Annotated lists of amphibians, reptiles, birds, and mammals of the region were also included. This was the beginning of the *life-zone system* which was developed by Merriam between 1890 and 1910 and continued by many others.[11] It continues to be used by biologists who live and work in Western North America.[12]

The studies noted are representative. There are many other excellent papers.[13] Probably its straightforward simplicity and verifiability are principal reasons for the vitality of the life-zone system and its continued use in the West by professional and non-professional naturalists alike (Jenks, 1931; Brandt, 1951; Heald, 1951; Arnberger, 1952; Patraw, 1953; Olin, 1959, 1961).

The number of classified zones is few (six in Arizona; Table 3) and they refer to obvious, easily observed vertical and latitudinal zones of plants and associated animals. Today, just as yesterday, these zones are based on the actually observable ecologic distribution of plants and animals and are mapped on the basis of the vegetation. They were not originally conceived (Merriam, 1890) nor followed (*e.g.*, Hall and Grinnell, 1919) and are not now recognized on the basis of temperature, moisture, or other physical factor(s), or "fauna" alone, contrary to what some would lead others to believe (see Allee, 1926; Shelford, 1932, 1945; Kendeigh, 1932, 1954, 1961).[14] The most competent, proper and

---

[11] For example, Stejneger (1893), Townsend (1893), Miller (1895, 1897), Cockerell (1897, 1898, 1900), Osgood (1900), Bailey (1902), Brown (1903), Bailey (1913, 1926, 1929, 1931, 1935, 1936), Grinnell (1908, 1928, 1935), Smith (1908), American Ornithologists' Union (1910), Cary (1911, 1917), Wooton and Standley (1913), Swarth (1914, 1920), Rydberg (1916), Howell (1917), Hall and Grinnell (1919), Johnson (1919), Howell (1921, 1938), Saunders (1921), Nelson (1922), Preble and McAtee (1923), Grinnell and Storer (1924), Jepson (1925), Grinnell, Dixon and Linsdale (1930), Mead (1930), van Rossem (1931, 1932, 1936a, b, 1945), McHenry (1932, 1934), Pearson (1933), Willett (1933).

[12] For example, Hargrave (1933a, b, 1936), Lutz (1934), Garth (1935, 1950), Vorhies, Jenks and Phillips (1935), Dodge (1936, 1938), Jones (1936, 1938), Gloyd (1937), Grater (1937), Graham (1937), Miller (1937, 1946, 1951), Cockerell (1941), Cooke (1941), Monson and Phillips (1941), Peterson (1941, 1942), Ball, Tinkham, Flock, and Vorhies (1942), Huey (1942), Monson (1942), Sutton and Phillips (1942), Aldrich and Friedman (1943), Hall (1946), Soper (1946), Ingles (1947), Dalquest (1948), Johnson, *et al.* (1948), Stebbins (1949, 1954), Little (1950), Murie (1951), Durrant (1952), Musebeck and Krombein (1952), Hoffmeister (1955), Cockrum (1955, 1963), Haskell (1955), Baker (1956), Castetter (1956), Van Tyne and Berger (1959), Phillips and Monson (1963).

[13] In investigations of the Olympic Peninsula and Mount Rainier, Jones (1936, 1938) determined the zonal percentages of the life-form classes in the Raunkiaeran system (see Cain, 1950).

Table 3. Temperatures (°F) in Biotic Communities in Northern Arizona in Winter (January) and in Summer (July). Data from Pearson (1920) for 1917 and 1918, and from U.S. Climatological Data (1917, 1918) for Needles, Calif.

Biotic Community	Mean Max.		Mean		Mean Min.		Minimum 1918	
	Jan.	July	Jan.	July	Jan.	July	Jan.	July
Desert								
Southern Desertscrub (Needles, 480 ft.) Lower Sonoran Life-zone	—	—	47	93	—	—	28	64
Grassland								
Desert-Grassland (Kingman, 3,300 ft.) Upper Sonoran Life-zone	54	98	41	83	28	67	21	53
Woodland								
Juniper-pinyon Woodland (Ash Fork, 5,100 ft.) Upper Sonoran Life-zone	49	90	34	74	20	59	0	50
Forest								
Ponderosa pine Forest (Flagstaff 6,900 ft.) Transition Life-zone	38	78	24	65	9	51	−13	42
Douglas fir Forest (San Francisco Mt., 9,100 ft.) Canadian Life-zone	30	68	25	58	19	49	−2	43
Spruce-fir Forest (San Francisco Mt., (10,700 ft.) Hudsonian Life-zone	26	59	21	53	16	46	−6	39
Tundra								
Alpine tundra (Timberline, 11,500 ft.) Arctic-Alpine Life-zone	22	55	16	48	10	41	−3	34

unbiased criticism of Merriam and the life-zone system has been that of Daubenmire (1938). The most useful discussion of relationships between, and uses of, zone, biome, province, etc., may be found in the work of Alden Miller (1951).

It is shown in Table 3 and Table 7 that some life-zones include more than a single biotic community; that is, a life-zone may be "generic" in being a higher level of classification including two or more biotic communities. In fact, the Upper Sonoran Zone includes too much (primarily woodlands and grasslands). It should be noted, however, that while a life-zone may be equal to a major biotic community, or contain two or more major biotic communities, a life-zone does not (in the Far West) artificially cut across natural biotic communities.

The life-zone system is not of utility on a world-wide or even continental scale, and this is only one of its shortcomings from philosophic and evolutionary points of view. It is imperfect, as are all ecologic classifications. Nevertheless, throughout subcontinental western North America it has provided a serviceable basis for faunal analysis and as the result of the soundness of the work of early naturalists, it has been found useful over a period of time equal to the age of American ecology itself, *i.e.*, since just before the turn of the century. As a tool, if not a completely satisfactory biogeographic system, it still endures; in short, it is simple, straightforward, and works throughout western North America.

Thus the authors of the Arizona vertebrate check lists, all of whom have studied over western landscapes for many years, are among those who are quick to point to the frequent usefulness of the life-zone system in the study of animals in the Southwest, and they use life-zone annotations. For western teachers in general, and especially for the elementary through high school grades, one can recommend the "life-zones" as a simple additional means to foster interest and understanding in the out-of-doors.

The following outline and discussion of the Merriam Life-zone System refers to the zones and their characteristics (*i.e.*, their biotic communities) as they occur in Arizona.

Lower Sonoran Life-zone

The Lower Sonoran Zone is equivalent to desert (Fig. 3-18). Parts of the three southern subdivisions of the North American Desert

---

[14] Merriam's subsequent and incorrect temperature facade (1894, 1899b), to explain control of the life-zones on an essentially one-factor basis, in no way vitiates the actual zonal distribution of plants and animals in nature, which observation, after all, is the basis for the concept of biotic belts or zones anywhere in the world where they occur. Actually, as was recently discussed elsewhere (Lowe, 1961), there is no need for either term (life-zone or biome) in the investigation of biotic communities and biogeography: the former is partly and the latter is wholly synonymous with *ecologic formation*. However, either term may be considered desirable for serving some purpose, as is the use of life-zone in the present check lists of vertebrates.

are represented in Arizona. These are the Mohave Desert, Sonoran Desert, and Chihuahuan Desert (the fourth subdivision is the more elevated Great Basin Desert which is treated in the Upper Sonoran Zone). The vegetation of all three may be collectively called the *Southwestern Desertscrub*, and that of the Great Basin Desert called the *Great Basin Desertscrub* (Fig. 26-29). A practically continuous belt of pine and juniper vegetation across the central part of the state effectively separates the two, the so-called "cold desert" (Great Basin Desertscrub) and "hot desert" (Southwestern Desertscrub).

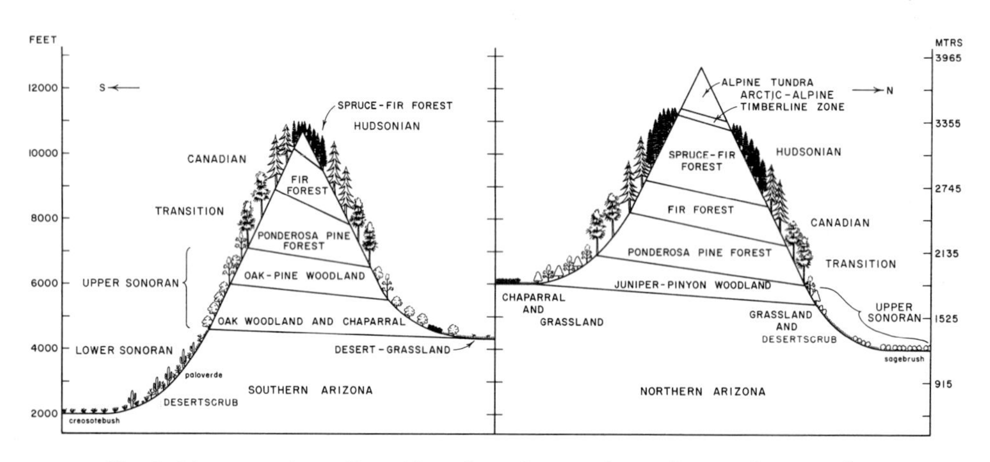

Fig. 2. Diagrammatic profiles of hypothetical mountains, indicating the vertical zonation of biotic communities in southern Arizona (left) and in northern Arizona (right). See Table 3, and text, for the corresponding life-zones.

The elevation of the zone is from 100 feet to 3,500-4,000 feet according to slope exposure; all of the life-zones extend to higher elevations on south-facing slopes than on north-facing slopes (Fig. 2). Precipitation (average annual) varies from approximately 3 to 11 inches; to about 12 inches in parts of the Chihuahuan Desert in the elevated southeastern corner of the state. It is distributed (1) primarily in winter (Mohave Desert; and Sonoran Desert, western part), or (2) more bi-seasonally (Sonoran Desert, eastern part) with somewhat more or less than half of it falling during the southwestern summer monsoon, or (3) primarily in summer (Chihuahuan Desert), with about three-quarters of the total yearly precipitation falling during the monsoon.

Desert vegetation often merges gradually, even imperceptibly, with desert-border vegetation and usually no real "line" of demarcation exists. This is particularly true, for example, in western Mexico, where there is no boundary "line" in what is a gradual south-north transition from southern subtropical thornscrub ("thorn forest") into desertscrub at the southern edge of the Sonoran Desert, in extreme southern Sonora and Baja California. This is a complex and highly informative subcontinental south-north gradient involving two primary physical controls operating on each species more or less independently of the others: these are (1) the northerly (toward U. S.) reduction in the *summer* rainfall and the total precipitation, with (2) progressively lower and critical *winter* minimum temperature. Accordingly, there is a gradual latitudinal change in community composition on this thornscrub-desertscrub environmental gradient. It involves a northward reduction in the total number of subtropical species of plants and animals as well as the amount of ground cover of the total vegetation—a progressive northward change in "openness" of the habitat, into the open and shorter-statured vegetation of the Sonoran Desert that is finally seen in southern Arizona and northwestern Sonora (Shreve, 1934).

Annual wildflowers (ephemerals) may in some years be one of the most conspicuous and colorful aspects of desert landscapes. It is well known that spring annuals (February-April) in southern Arizona may be remarkably abundant following good winter rains (from the northwest) and the plants are largely "Californian" species of mustards, poppies, sand-verbenas, etc., whereas ephemeral flower displays during July-September following good summer rains from the south are largely of "Mexican" aspect — amaranths, morning glorys, zygophylls, *et al.* Thus as one travels across the North American Desert from north to south or west to east — say, from Nevada to Sonora or from California to Chihuahua — there is an increase that one would expect in the "Mexican" component of the desert annual flower displays, just as there is in the composition of the desert perennial trees and shrubs. The plant adaptation thereby expressed is correlated with the underlying NW to SE climatic gradient away from the primarily winter (cool) precipitation of the Californian area to the predominantly summer (warm) rainfall in the Sierra Madrean area (Shreve, 1944).

It is of further interest that the more primitive relatives of many desert *annuals* are tropical and subtropical *perennial* herbs, shrubs, and trees; for example, in the families Amaranthaceae, Convolvulaceae, Hydrophyllaceae, Leguminosae, and Zygophyllaceae. This is one of numerous neontological, in addition to paleontological, relationships which are consistent with the evolutionary fact that the desert is our youngest major earth environment (see Axelrod, 1950; Shreve, 1951). And, moreover, that the ephemeral growth form (i.e., the plant life-form of holding over only as earth-covered seed) is a particularly successful one for life in harsh environments such as deserts — this is drought-evasion par excellence.

The deserts have evolved through grassland stages to the present desert conditions in response to increasing aridity since Mio-Pliocene time, and some "tougher" species of grasses still occur naturally in some of the harshest desert environments. Thus grasses may seem virtually absent over extensive areas of desert, or they may be obviously plentiful under more local conditions. Several species of *Aristida* (three-awn) are more or less successful throughout Southwestern desertscrub in Arizona.[15] The small compact desert fluff grass (*Tridens pulchellus*) in the open, and bush muhly (*Muhlenbergia porteri*) clumped around bases of shrubs, are examples of common species widely distributed but not always conspicuous throughout much of the desert. Other grasses such as big galleta (*Hilaria rigida*) and tobosa (*Hilaria mutica*) may cover conspicuously well-developed local grassland swales, which microenvironments persist well within the desert (Fig. 26). Also, on dry rocky hillsides of desert ranges which are steeply sloped, grasses occasionally may be in fair abundance as is the case with curly mesquite grass (*Hilaria belangeri*) on the Tucson Mountains and similar low ranges and desert mesas in the southern part of the state. Or, by sharp contrast, they may be next to non-existent, as are the few clumps of tanglehead grass (*Heteropogon contortus*) existing on the rugged and nearly barren volcanic desert hills in the Boulder Dam area of the Colorado River in the northwestern corner of the state.

Chihuahuan Desert

The Chihuahuan Desert is represented in Arizona by small and sometimes isolated areas in the southeastern corner of the state, primarily in Cochise County (Fig. 3-5). Parts of San Simon Valley, Sulphur Springs Valley, and San Pedro Valley have well-developed communities dominated by tarbush (*Flourensia cernua*), creosotebush (*Larrea divaricata*), sandpaperbush (*Mortonia scabrella*), or Chihuahuan white-thorn (*Acacia constrictor vernicosa),* which comprise the four major association-types of this desert occurring in Arizona. In many areas tarbush, creosotebush, and white-thorn are intermixed and associated with such species as allthorn (*Koeberlinia spinosa*), desert sumac (*Rhus microphylla*), shrubby senna (*Cassia wislizeni*), ocotillo (*Fouquieria splendens*), mesquite (*Prosopis juliflora*), and others (Fig. 3). With the exception of creosotebush, ocotillo, and mesquite (which have wide ranges in both the Chihuahuan and Sonoran Deserts, and also occur in the Mohave Desert), these are species among a small group of plants which have wider distributions in the Chihuhuan Desert in northern Mexico and southern New Mexico, and enter only the southeastern corner of Arizona at the northwestern limits of their ranges.

The Chihuahuan Desert is essentially a shrub desert as is the Great Basin Desert. It lies mostly above 3,500 feet in elevation on the "Mexican

---

[15] In the Great Basin desertscrub scattered colonies or large populations of galleta (*Hilaria jamesi*), blue grama (*Bouteloua gracilis*), sacaton (*Sporobolus wrighti*), and three-awns (*Aristida*) are not uncommon.

Fig. 3. Southwestern Desertscrub, Chihuahuan Desert. Lower Sonoran. Looking south toward Naco Mountain, Sonora, from 4 miles NW of Naco, 4,600 ft., Cochise County. Tarbush (*Flourensia cernua,* left foreground), creosotebush (*Larrea divaricata,* right center), and white-thorn (*Acacia constricta,* small individual right foreground), with Thornber yucca (*Yucca baccata thornberi*) and grasses — a relatively mixed stand of these species, common on level valley fill.

Plateau," and the Great Basin Desert lies mostly above 4,000 feet. While it comprises small areas of Arizona, New Mexico, and Texas, it lies primarily in the states of Chihuahua and Coahuila and in parts of Durango, Zacatecas, Neuvo Leon, and San Louis Potosi. Generally lower elevations are reached in Trans-Pecos, Texas, and adjacent Chihuahua where the Chihuahuan Desert borders the Rio Grande to as low as approximately 1800 feet elevation. In several parts of Cochise County, Arizona, and in adjacent New Mexico, Western Texas, Chihuahua, and Sonora, the desert and grassland vegetations are found in complex mixtures. Desert and grassland species form interesting landscape mosaics controlled importantly by marked changes in soil conditions over very short distances.

Fig. 4. Southwestern Desertscrub, Chihuahuan Desert. Lower Sonoran. Looking northwest across San Pedro Valley, 3.5 miles E of Fairbank, Cochise County. Tarbush, white-thorn, and creosotebush dominate the rolling, calcareous-soiled landscape in stands of various mixtures which include other shrubs and cacti: ocotillo (*Fouqueria splendens*), mariola (*Parthenium incanum*), allthorn (*Koeberlinia spinosa*), Mexican crucillo (*Condalia spathulata*), and "chollas" including Christmas cactus (*Opuntia leptocaulis*).

Grasses are more abundant in the more highly elevated Chihuahuan Desert than in the Sonoran Desert. In fact, some investigators have considered all of the Chihuahuan Desert to be a grassland climax (Weaver and Clements, 1938; Whitfield and Beutner, 1938; Whitfield and Anderson, 1938; Gardner, 1951). However it may have been in the distant past, today the Chihuahuan Desert is definitely a natural desert dominated by a host of climax shrubs (Shreve, 1917, 1939, 1942a, 1942b, 1942c; Livingston and Shreve, 1921; Shantz and Zon, 1924; Muller, 1939, 1940; Benson and Darrow, 1944; Leopold, 1950; Rzedowski, 1956). The problem is confused by the recent (historical) environmental changes and desert shrub invasions, particularly along the northern edge of the desert in the vicinity of the international boundary where the grassland and desertscrub have long met to form a mosaic landscape pattern (Shreve, 1939, 1942c; Gardner, 1951; Ditmer, 1951; Lowe, 1955; Yang, 1961).

The problem was closely examined and clearly explained by Shreve (1942c, and elsewhere). The so-called "Desert Grassland" is not merely

Fig. 5. Southwestern Desertscrub, Chihuahuan Desert. Lower Sonoran. Sandpaper-bush (*Mortonia scabrella*) in an extensive and essentially pure stand on limestone soil, Cochise County. Foreground is east edge of the town of Tombstone, 4,500 ft., looking northeastward toward Dragoon Mountains. The general appearance of this desertscrub landscape is superficially like that of chaparral.

a grassland; it is obvious also that it is not desert plains and that the term "desert plains grassland" is a poor one. This is a broad and highly varied transition region between the plains grassland (short-grass plains) and the Southwestern Desertscrub of more recent evolution (Axelrod, 1950). The climate is intermediate between desert and grassland, and a slight change in the precipitation-evaporation ratio (for example, by a slight but significant rise in evironmental temperature) can effect a pronounced change in the vegetation at a given locality.

This is happening at middle latitudes in western North America, and elsewhere, today. A small but significant climatic change toward somewhat warmer and drier conditions is "written" into diverse records (see Kincer, 1946; McDonald, 1956; Schulman, 1956).

This complex situation is to be seen on many parts of the range ecologists "beat" which lies just north of the international boundary and between southcentral Texas and southcentral Arizona, an area which has been repeatedly examined. It should be reiterated at this point that the Chihuahuan Desert lies essentially south of this transect to beyond the state of Chihuahua and into parts of Zacatecas, Durango, Nuevo Leon, and San Luis Potosi; roughly 90 per cent of it is in Mexico. Not more than a handful of ecologists have examined its southerly parts extensively or intensively; the most notable of these have been Shreve, Muller, and Rzedowski. The idea that none of the Chihuahuan Desert is desert, and that all of it is subclimax grassland (references above), is to be emphatically rejected.

Two important papers dealing with genetic variation, hybridization, and post-Pleistocene recontact of animal populations in the desert-grassland transition have been recently published (Zweifel, 1962; Dessauer, Fox, and Pough, 1962). The region investigated in southern Arizona-New Mexico is in part a complex continuum as well as mosaic transition between biotic components of the Chihuahuan Desert, Plains Grassland, and Sonoran Desert.

The Chihuahuan Desert in Arizona has a relatively poor representation of cactus life-forms and species when it is compared with the Sonoran and Mohave Deserts, but is a richer region in this regard than is the more northerly and cooler Great Basin Desert. Chollas, prickly pears, barrels, and pincushions are represented in this southeastern region of Arizona by the following more common species: cane cholla, *Opuntia spinosior;* desert Christmas cactus, *O. leptocaulis;* devil cholla, *O. stanlyi;* shrubby prickly pear, *O. macrocentra;* Engelmann prickly pear, *O. engelmanni;* Wislizenius barrel cactus, *Echinocactus wislizeni;* common pincushion, *Mammallaria vivipara aggregata;* and devil's pincushion, *Mammillaria robustispina.*

Outside of Arizona, and particularly in Mexico, there is a considerable number of other low-growing cacti, leaf succulents (*e.g.*, agaves), and stem semisucculents (yuccas) in the Chihuahuan Desert and desert-grassland. Lechugilla (*Agave lecheguilla*) and Torrey yucca (*Yucca torreyi*) are examples of these which occur as far north as southern (SW) Texas and New Mexico but which do not reach westward to Arizona.

Sonoran Desert

The Sonoran Desert covers most of southwestern Arizona. It is the hottest of our southwestern deserts, and the lowest elevation point in the State—*ca.* 100 feet on the Colorado River in "Yuma Valley"—is in this desert region. Rainfall is distinctly bi-seasonal (Arizona upland section) or occurs primarily during the winter (lower Colorado section). In Arizona the Sonoran Desert is most widely characterized by two of its principal terrestrial biotic communities, the "creosotebush communities" and the "paloverde communities."

In the paloverde-sahuaro (*Cercidium-Cereus*) community (Fig. 6, 7, 9), the plants are comprised of small-leaved desert *trees* as well as of shrubs and numerous cacti, and best development is attained on rocky hills, bajadas, and other coarse-soiled slopes in the succulent Arizona upland desert section, *e.g.*, between Ajo and Tucson. It is a particularly rich

Fig. 6. Southwestern Desertscrub. Lower Sonoran. Paloverde-sahuaro community on granite rock in Arizona Upland section of Sonoran Desert; Shreve's Ocelada Camp in Soldier Canyon, *ca.* 3000 ft., south side Santa Catalina Mountains, Pima County. In addition to foothill paloverde (*Cercidium microphyllum*) and sahuaro (*Cereus giganteus*), teddy bear cholla (*Opuntia bigelovi*), ocotillo (*Fouquieria splendens*), and brittlebush (*Encelia farinosa*) are important species in the community, particularly on south-facing slopes. Photo by Forrest Shreve, 1915. Today the Mt. Lemmon (Catalina) Highway winds upward from Soldier Trail junction, near lower right corner of picture. Few and relatively minor changes in the natural vegetation have taken place here during the past half-century.

community of desert plants and animals exhibiting highly varied and often spinose life-forms. Shrubs are more varied than the trees, and while the foothill understory may be predominantly of a single species such as triangle bur-sage (*Franseria deltoidea*) or brittlebush (*Encelia farinosa*), it is often comprised of a mixture of 5 to 15 or more shrub and dwarf shrub species in the form of a three, four or five layered understory. The primary desert (non-riparian) trees are foothill paloverde (*Cercidium microphyllum*), sahuaro (*Cereus giganteus*), ironwood (*Olneya tesota*), elephant tree (*Bursera microphylla*), crucifixion thorn (*Canotia holocantha*), desert-olive (*Forestiera phillyreoides*), holocantha (*Holocantha emoryi*),

Fig. 7. Southwestern Desertscrub, Sonoran Desert. Lower Sonoran. At 2,800 ft. near pass into Sabino Canyon, Santa Catalina Mountains. The substratum supporting the foothill paloverde community is typically rocky, being on and near the parent bedrock source of raw soil material. Life-form is highly diverse and community biomass and productivity are greater here than in the creosotebush communities on the valley fill of the plains below. This tree-dominated desert climax is not successional to, nor succeeded by, the shrub-dominated creosotebush climax desert community or any other. Photo by Forrest Shreve.

the tree-like chollas (particularly *Opuntia fulgida*), and the columnar organ-pipe cactus (*Cereus thurberi*) and senita (*Cereus schotti*).

The primarily desert *riparian* trees of the "dry" arroyos and washes include blue paloverde (*Cercidum floridum*), mesquite (*Prosopis juliflora*), catclaw (*Acacia greggi*), smoketree (*Dalea spinosa*), desert willow (*Chilopsis linearis*), jumping bean (*Sapium biloculare*), and netleaf hackberry (*Celtis reticulata*); these trees and several associated shrubs provide often well-developed riparian associations far within the desert where cottonwoods and willows (riparian woodland) do not penetrate. Such desert riparian associations are nearly as prevalent among the creosotebush as well as paloverde communities, and in the Yuma area even sahuaros, foothill paloverdes, and ironwoods are found in such arroyo habitats.

Fig. 8. Southwestern Desertscrub, Sonoran Desert. Lower Sonoran. The larger (*Larrea divaricata*) and the smaller (*Franseria dumosa*) shrubs of the creosotebush-bur-sage community on sandy soil, with no other perennial plants, as is common over extensive areas of both the Mohave and Sonoran Deserts. Yuma County, on the Yuma-San Luis Mesa, 175 ft., 7 miles E of San Luis, Arizona.

The other of the two major plant-animal communities is the much simpler creosotebush-bur sage (*Larrea-Franseria*) community (Fig. 8) which is composed mainly of *shrubs* and dwarf shrubs. Over great areas the plant dominants are essentially creosotebush (*Larrea divaricata*) and white bur-sage (*Franseria dumosa*), growing either together or alone (Fig. 10, 11). Other occasional dominants are mostly shrubs. Trees are usually lacking except for those which are desert riparian trees in the drainageways (as noted above). This shrub community characterizes habitats less rocky and of lower relief, such as valleys, mesas, and shelving plains in the lower Colorado desert section of the Sonoran Desert (the so-called "Colorado Desert"). There is often interspersion of these two major climax communities and their habitats across southwestern Arizona as may be readily observed along the highway between Yuma and Tucson, Blythe and Phoenix.

In addition to the paloverde and creosotebush associations, desert saltbush (*Atriplex polycarpa*) frequently forms extensive stands across valley bottomlands which are periodically flooded and which have soils of

Fig. 9. Southwestern Desertscrub, Sonoran Desert. Lower Sonoran. Organpipe cactus (*Cereus thurberi*) in the foothill paloverde community in Alamo Canyon, 2,400 ft., Ajo Mountains, Organ Pipe Cactus National Monument, Pima County. Triangle bur-sage (*Franseria deltoidea*) dominates the foreground.

Fig. 10. Southwestern Desertscrub, Sonoran Desert. Lower Sonoran. Pure stand of creosotebush (*Larrea divaricata*) on deep sand of the Yuma-San Luis Mesa, 150 ft., Yuma County, a few feet north of the Arizona-Sonora boundary fence, 4 miles E of San Luis del Rio Colorado, Sonora.

Fig. 11. Southwestern Desertscrub, Sonoran Desert. Lower Sonoran. Pure stand of white bur-sage (*Franseria dumosa*) on sand, one-half mile N of Dateland, 450 ft., Yuma County. This is a small stand about 300 yards diameter within the surrounding creosotebush-bur-sage community of San Cristobal Valley. Low, stabilized sand dunes on horizon are capped with western honey mesquite (*Prosopis juliflora torreyana*); dark spots in mesquites are clumps of mistletoe.

fine texture that are more or less alkaline, and saltbush (*Atriplex polycarpa*, and others) may be considered an important vegetation type in the Sonoran Desert. Mesquite (*Prosopis* spp.) formerly grew along many of the larger desert drainageways, such as those along the Gila River and some of its tributaries, in dense forest-like stands called mesquite bosques. These valuable riparian trees reached heights of 40 to 50 feet, with some individuals having trunks 2 to 3 feet in diameter. Remnants of such stands are still present in scattered areas along San Pedro River and Santa Cruz River as well as parts of the Gila and lower Colorado, and many smaller bosques still remain. In addition to the four major kinds of association-types dominated primarily by foothill paloverde, creosotebush, saltbush, and mesquite, some minor association-types in the Sonoran Desert in Arizona are dominated locally by other species such as jojoba (*Simmondsia chinensis*) on rocky upland sites within the paloverde type, white bursage (*Franseria dumosa*) within the creosotebush type, and ocotillo on elevated, rocky shallow soils; small local areas of almost pure stands of these, and some other species are occasionally encountered (Fig. 11).

Fig. 12. Southwestern Desertscrub, Sonoran Desert. Lower Sonoran. A desert wash, the usually dry bed and floodplain of the Hassayampa River at Gates Ranch, *ca.* 1750 ft., near Morristown, Maricopa County. The mesquite-acacia floodplain (mesquite bosque) is dominated by velvet mesquite (*Prosopis juliflora vetulina*). The climax of the surrounding rocky hills is paloverde-sahuaro and that of the non-riparian valley plains is creosotebush-bursage. Photo by Gerald O. Gates and E. Curtis Arnett.

Along the formerly great Gila River (the now dry bed of which stretches across the Sonoran Desert of western Arizona) there were extensive marshes, swamps, and floodplains with cattail (*Typha domingensis*), bulrush (*Scirpus olneyi*), giant reed (*Arundo donax*), common reed (*Phragmites communis*), arrowweed (*Pluchea sericea*), and many trees. The dense vegetation of these well-developed riparian communities often stood 10 to 15 feet high and supported a tremendous quantity and variety of wildlife. Today such habitats persist in modified form along the lower Colorado River and along parts of the greatly changed Gila where its remnant persists in east-central Arizona; tamarix (*Tamarix*) is an increasingly abundant foreign introduction in some of these riparian situations and may become locally undesirable.

The Sonoran Desert contains the most diverse cactus flora in our Southwest. Many species representing all of the southwestern cactus life-forms are present: columnar and giant cactus (*Cereus*), barrels (*Echino-*

Fig. 13. Southwestern Desertscrub, Sonoran Desert. Lower Sonoran. Creosotebush on rough desert pavement, 8 miles W of Sentinel, Yuma County. Plant in foreground is partly dead and 45 per cent of the stand is completely dead today on this black, rocky surface under 3-4 inches annual rainfall. No plant succession takes place here; the same species and often the same individuals are both the pioneer and the climax plants.

cactus), chollas (*Opuntia*), prickly pears (*Opuntia*), hedgehogs and rainbows (*Echinocereus*), and the pincushions and fish hooks (*Mammillaria*). In order of decreasing richness of cacti in the flora, our deserts are: Sonoran, Mohave, Chihuahuan, Great Basin.

Mohave Desert

The Mohave Desert is a transitional area between the more highly elevated and cooler Great Basin Desert in the north and the hotter Sonoran Desert in the south. In Arizona this desert is in the western and northwestern part of the state, northward of the approximate line Needles — Congress Junction, and into the extreme northwestern corner within and north of the valley of the Grand Wash Cliffs, to (and across) the Utah state line. It covers considerably greater desert area in southeastern California, and also occurs in southern Nevada and in extreme southwestern Utah, in the Virgin River drainage to a few miles north of St. George and east of Hurricane. Most of the dominant plant species have their main distribution in the other deserts and creosotebush is the primary

Fig. 14. Southwestern Desertscrub, southern edge of Mohave Desert-Sonoran Desert transition, 12 miles NW of Congress Junction, *ca.* 3,000 ft., Yavapai County, looking eastward toward Date Creek Mountains. Lower Sonoran. Joshuatree (*Yucca brevifolia*) is a Mohave species. Foothill paloverde (*Cercidium microphyllum*), tree on left, is a Sonoran one. Creosotebush, left-center edge and across center background, is the most abundant and conspicuous plant species common to both deserts.

shrub. Open stands of creosotebush and white bur-sage occur more or less throughout. Only three plants usually stand above the generally low shrubs: Joshuatree and Mohave yucca at the somewhat higher elevations, and catclaw along washes.

The principal association-types in the Mohave Desert in Arizona are creosotebush, Joshuatree, blackbush, and saltbush; also bladder-sage. Bladder-sage (*Salazaria mexicana*) is a widespread and nearly endemic species of the Mohave Desert and in some areas it may form nearly pure stands. Other and more minor associations on the Mohave are often in transitional areas between major associations. These may be partly to almost wholly dominated by white bur-sage, shrubby buckwheat (*Eriogonum wrighti*), other shrubs (such as snakeweed) and some grasses such as big galleta (*Hilaria rigida*) and tobosa (*H. mutica*).

One of the best-known life-forms is the Joshuatree (*Yucca brevifolia*) which forms often extensive "forests" or "woodlands" of varied associations. The understory vegetation may be dominated by a shrub species of the northern desertscrub (*e.g.*, blackbrush, Fig. 17), or by one of the southern desertscrub (*e.g.*, creosotebush), or by a species more or less endemic to

Fig. 15. Southwestern Desertscrub, Mohave Desert. Lower Sonoran. Creosotebush-bur-sage with Mohave yucca (*Yucca schidigera*) and other shrubs in Detrital Valley, Mohave County. Looking westward toward Black Mountains, 21 miles NW of Chloride, *ca.* 2,400 ft.

the Mohave desert area (*e.g.*, Parish devil cholla, Fig. 18). More often, however, the plants associated with the Joshuatree are several species in various complex mixtures, with mostly Great Basin species in the northern part of the Mohave Desert and Sonoran Desert species in the southern part.

At the southern limit of the Joshuatree's distribution in Arizona, a few miles northwest of Congress Junction, Yavapai County (Fig. 14), it grows adjacent to and often in association with species of the Sonoran Desert such as foothill paloverde, sahuaro, ocotillo, and crucifixion thorn, as well as creosotebush and white bur-sage. Here, in a transitional area of Mohave desert and Sonoran desert, there are no typically Great Basin desert species (*e.g.*, sagebrush, shadscale) associated with Joshuatree.

The Mohave yucca (*Yucca schidigera*, Fig. 15) is also conspicuous over large areas of the Mohave desert where it is commonly in association with creosotebush, white bur-sage, bladder-sage (*Salazaria mexicana*), brittlebush (*Encelia farinosa*), chollas, barrels, etc. The ecologic distribution of Mohave yucca overlaps that of the Joshuatree, and while both species frequently are together, as on the upper parts of outwash slopes along the bases of desert ranges, the Mohave yucca also occurs at some-

Fig. 16. Southwestern Desertscrub, Mohave Desert. Lower Sonoran. Small trees of catclaw (*Acacia greggi*) in Detrital Wash, Detrital Valley, *ca.* 2,400 ft., Mohave County. Catclaw is one of the few desert riparian trees in the Mohave Desert, and the only one commonly seen throughout the Mohave in Arizona. It grows as a shrub in non-riparian desert habitats.

what lower elevations. Banana yucca (*Yucca baccata*) also occurs with both Mohave yucca and Joshuatree, and the three yuccas occur together at many localities in Mohave County.

Often in association with one or more of these yuccas, several life-forms of cacti are conspicuous, although cacti on the Mohave desert are by no means as varied or as numerous as in the Sonoran Desert and are largely restricted to the coarse soils on the gentle outwash slopes. These include chollas (*e.g.*, buckhorn cholla, *Opuntia acanthocarpa;* Mohave cholla, *O. echinocarpa;* shrubby cholla, *O. ramosissima;* Parish devil cholla, *O. stanlyi parishi*), prickly pears (*e.g.*, Mohave prickly pear, *Opuntia erinacea;* beaver tail cactus, *O. basilaris;* variable prickly pear, *O. phaeacantha*), hedgehog cactus (*Echinocereus engelmanni*), barrel cactus (*Echinocactus acanthodes*), and the desert pincushion (*Mammillaria vivipara deserti*). Also the sahuaro (*Cereus giganteus*) reaches its

Fig. 17. Southwestern Desertscrub, Mohave Desert, Lower Sonoran. Joshuatree "forest" (*Yucca brevifolia*) with a dense stand of blackbrush (*Coleogyne ramosissima*) interspersed primarily with banana yucca (*Yucca baccata*), *ca.* 3,400 ft., on Pierce Ferry road looking eastward toward Grand Wash Cliffs, Mohave County.

northernmost limit in the southern part of the Mohave desert, west of the Hualapai Mountains in southern Mohave County.

One of the striking differences between the Mohave desert (also Great Basin desert) and the Sonoran desert is the relative paucity of desert trees in the Mohave. Even along large arroyos and other drainageways in Mohave they are in near absence, both in kinds and in numbers of individuals. Three southerly riparian species of trees are present, however spotty in distribution, in the Mohave in Arizona: desert willow (*Chilopsis linearis*), western honey mesquite (*Prosopis juliflora torreyana*), and catclaw (*Acacia greggi*). Catclaw, either as a small tree or shrub, is the species more commonly seen throughout most of the Mohave (Fig. 16), particularly in Arizona.

Fig. 18. Southwestern Desertscrub, Mohave Desert. Lower Sonoran. Open Joshua-tree "woodland" on the Pierce Ferry Road, *ca.* 2,500 ft., Mohave County, with a conspicuous dwarf cactus "ground cover" of Parish devil cholla (*Opuntia stanlyi parishi*) interspersed with occasional creosotebush and banana yucca.

Upper Sonoran Life-zone

The Upper Sonoran Zone includes *woodland* (Fig. 33-39), *chaparral* (Fig. 32), *grassland* (Fig. 23-29) and *Great Basin desertscrub* (Fig. 19-22). Two exceptions to this "inclusion" should be noted, as follows: (1) riparian woodland (Fig. 40, 64-66), which occurs throughout all of the life-zones except the highest two, Hudsonian and Arctic-Alpine, and (2) mountain grassland (Fig. 31) which occurs in (forest) zones above the Upper Sonoran. Elevation of the zone is 3,500 to 4,000 feet to as high as 7,000 feet on some slopes. Precipitation varies from as little as 7 or 8 inches in the shrub-dominated Great Basin Desert to as much as 21 or 22 inches in tree-dominated woodland communities.

Great Basin Desert

The southeastern limit of the Great Basin desertscrub lies in the northern part of Arizona, principally in the region north and east of Flagstaff. It also occurs in areas in the extreme northwest, near the Utah State line, being particularly well represented in the Strip Country north of the Grand Canyon. This is the most highly elevated of the four deserts

Fig. 19. Great Basin Desertscrub, Great Basin Desert. Upper Sonoran. Basin sagebrush (*Artemisia tridentata*) in the Arizona Strip country north of the Colorado River. The pure stand of sagebrush is bordered on the distant horizon by juniper-pinyon woodland. Photo by Robert R. Humphrey.

(approximately 3,000 to 6,500 feet) with most of it occurring above 4,000 feet. Accordingly, it is also the coolest and it is sometimes called the cool desert, cold desert, semi-desert, etc. Precipitation is more evenly distributed throughout the year than it is in the other desert regions of the state, and is approximately 7 to 12 inches annually.

The Great Basin is a shrub (and grass) dominated desert in which the vegetation is of relatively low stature (Fig. 19-22) and is more or less uniform, with just a few species (and often only one) comprising the stand over extensive areas of similarly uniform relief. Trees are almost totally absent, and the shrubs have small leaves which are wholly or partly deciduous. The landscapes are more often spoken of as monotonous; they are, nevertheless, interesting to those who inspect them closely.

Major shrubs are big sagebrush (*Artemisia tridentata*), blackbrush (*Coleogyne ramosissima*), shadscale (*Atriplex confertifolia*), mormon-tea (*Ephedra viridis* and others), and greasewood (*Sarcobatus vermiculatus*). Each of these shrubs often forms more or less pure stands, and commonly with little more than a few associated grasses present; sagebrush, blackbrush, and shadscale form the principal groups of associations in Arizona.

Fig. 20. Great Basin Desertscrub, Great Basin Desert. Upper Sonoran. A pure stand of shadscale (*Atriplex confertifolia*) near Fredonia, Coconino County, 5,000 ft. One of the principal shrubs of the southern part of the Great Basin Desert, shadscale covers extensive areas in essentially pure stands, as does sagebrush (*Artemisia*) and blackbrush (*Coleogyne*).

In some areas other shrubs are of occasional prominence, such as four-wing saltbush (*Atriplex canescens*), black sagebrush (*Artemisia nova*), sand sagebrush (*Artemisia filifolia*), rabbitbrush (*Chrysothamnus nauseosus*), snakeweed (*Gutierrezia sarothrae*), plateau yucca (*Yucca angustissima*), pale lycium (*Lycium pallidum*), desert olive (*Forestiera neomexicana*), and serviceberry (*Amelanchier utahensis*). Relatively few species of cacti occur in this desert in Arizona, and none appear restricted to it. The most abundant are prickly pears (*e.g.*, grizzly bear cactus, *Opuntia erinacea ursina;* Navajo prickly pear, *O. erinacea hystricina;* western prickly pear, *O. polyacantha;* fragile prickly pear, *O. fragilis*) and chollas (*e.g.*, Whipple cholla, *Opuntia whipplei*).

The region in northeastern Arizona between the Little Colorado River and the Hopi Mesas was originally named the "Painted Desert" by Geologist Newberry during Ives' (1861: 76-78) exploration of the Colorado River (Dellenbaugh, 1932; McKee, 1933).[16] The Painted Desert is a minor subdivision of the Great Basin Desert which lies along the Little Colorado River below approximately 5,000 feet (1,500 meters) in elevation as properly indicated by Merriam (1890) and by Sellers (1960). It lies in

[16] Dr. J. S. Newberry (1861) wrote the report on the geology of the Ives Expedition to the Colorado River of the West, which constitutes Part III of the Ives Report.

Fig. 21. Great Basin Desertscrub, Great Basin Desert. Upper Sonoran. On the Arizona-Utah border, looking northwest from Comb Ridge across Monument Valley, 5,200 ft., to Navajo Mountain (the dome in the distant horizon, right center). In the immediate foreground is an open juniper-pinyon woodland along the upper edge of the desert valley. Photo by E. Tad Nichols.

eastern Coconino County, roughly between Tuba City near the northern end and Leupp near the southern end. Thus the parts of the "Painted Desert" of recent popular writers that are actually desert (*e.g.*, Jaeger, see footnote 7) lie wholly within the Great Basin Desert. Among the few species of widely scattered plants which occur at variously spaced intervals on the mostly bare ground, the most common are saltbushes (shadscale and fourwing) and grasses (primarily the dropseed grass called sacaton). The bareness of the ground (*e.g.*, north of Cameron, Coconino county) contributes to the spectacular displays of color which give this area its name of "painted," for there is widespread and almost complete exposure of the brilliantly colored soil layers of the actively eroding hills of shale (siltsone, *et al.*) which were originally laid down as old lake deposits during Upper Triassic time many millions of years ago (Fig. 22).

Fig. 22. Great Basin Desertscrub, Great Basin Desert. Upper Sonoran. The characteristic Chinle marl of the "Painted Desert," at *ca.* 10 miles north of Cameron, 4,200 ft., Coconino County. The delicate, striking coloring of these barren eroding hills of shale led Geologist J. S. Newberry of the Ives expedition to give the name "painted desert" to this area between Leupp (northwest of Winslow) and Tuba City. This relatively small area within the Great Basin Desert lies essentially below 5,000 feet elevation in the immediate drainage of the Little Colorado River. Photo by E. Tad Nichols.

Grassland

Desert-Grassland. Desert-grassland (Fig. 23-29) is a transitional type of grass-dominated landscape commonly positioned between desert below and evergreen woodland or chaparral above. Its lower limit is about 3,500 feet elevation and its best development is between 4,000 and 5,000 feet. Most of the desert-grassland in Arizona receives 10 to 15 inches of precipitation annually; the extremes are about 9 and 18 inches with the mean lying somewhere between 12 and 15. In Arizona, it is largely contained in the southeastern quarter of the state, but also occurs in the northwestern quarter, as in the vicinity of Kingman in Mohave County (Fig. 25). The grasses are often bunch-growth perennials in which the bases of the clumps are separated by intervening bare ground.

Fig. 23. Desert-grassland. Upper Sonoran. A landscape of palmilla (*Yucca elata*) and grama grasses (*Bouteloua*) in Cochise County. This is the principal landscape (yucca-grass) which many Arizonans think of when the name "desert grassland" is used. However, the stand of grama shown here is more dense than is usual over most of the desert-grassland transition in southern Arizona today. Photo by Robert R. Humphrey.

Where soil may be deep, well-protected from erosion and with few rocks, shrubs, or cacti, perennial grama grasses such as black grama (*Bouteloua eriopoda*), blue grama (*B. gracilis*), sideoats grama (*B. curtipendula*), slender grama (*B. filiformis*), and hairy grama (*B. hirsuta*) may cover, even today, extensive stretches of landscape. Many other grasses are mixed with the gramas, *e.g.*, plains lovegrass (*Eragrostis intermedia*), plains bristlegrass (*Setaria machrostachya*), sand dropseed (*Sporobolus cryptandrus*), and cottongrass (*Trichachne californica*), and several species of three-awn (*Aristida*).

Such purely grass landscapes, however, stand in marked contrast to other desert-grassland cover on much of the shallow-soiled, rocky and gravelly hills and slopes which are of considerable extent in these parts of Arizona. On such shallower soils the climax grasses are usually much reduced and some of them eliminated in competition with a wide variety of shrub, tree, and cactus life-forms such as prickly pears and chollas, agaves, yuccas, ocotillo, fairy duster, wait-a-minute bush, cat-claw, and mesquite. Sotol (*Dasylirion wheeleri*) and beargrass (*Nolina microcarpa*) are shrub life-forms which are occasionally conspicuous and may even

Fig. 24. Desert-grassland. Upper Sonoran. Thornber yucca (*Yucca bacata thornberi*) in a nearly pure stand extends as far as the eye can see. A few widely scattered palmilla and sotol (*Dasylirion wheeleri*) also occur and the small-clumped fluff grass (*Tridens pulchellus*) is a predominant species. Looking southwestward toward Mount Fagan from near Mountain View (between Tucson and Benson), *ca.* 3,600 ft., Pima County.

dominate local situations on shallow soils. Only the tougher grasses (that are, incidentally, of lesser forage value) may be present or abundant, such as ring grass (*Muhlenbergia torreyi*), red three-awn (*Aristida longiseta*), and fluff-grass (*Tridens pulchellus*). Mesquite has invaded large areas of former grassland as have other trees (*e.g.*, juniper) and some shrubs both native and introduced. Investigations have included before-and-after photographs which are especially informative aids, *e.g.*, the photographs of Parker and Martin (1946) for grass stand in 1903 to mesquite stand in 1941 on the north side of the Santa Rita Mountains.

The extremes of the desert-grassland in Arizona have just been briefly noted, *i.e.*, (1) grass landscapes of essentially pure stands of grasses and (2) the other more frequent extreme of mixed grass-shrub landscapes, often expressed as shrub and/or tree stands with varying grass composition and depletion. Between these extremes, two very characteristic desert-grassland habitats in Arizona today are represented by (1) extensive yucca-grass landscapes on undulating terrain (Fig. 23, 24) and (2) tobosa

Fig. 25. Desert-grassland. Upper Sonoran. Looking northwestward, 3,400 ft., across Hualapai Valley toward Cerbat Mountains which extend northward from Kingman, Mohave County. The shrubs, yuccas, and cacti as well as the grasses in this desert-grassland in Mohave and Yavapai counties are often of the same genera and species as those of the similar desert-grassland in southeastern Arizona.

grass (*Hilaria mutica*) in flat valley bottomlands and swales (Fig. 26). The differences in plant life-form as well as plant and animal species which occur in these two characteristic kinds of southwestern biotic communities are the result of the differences in topography, run-off, and soil characteristics which prevail in an extensive habitat mosaic under the same macroclimate.

Plains Grassland. Well developed plains grassland, in which the grasses form a continuous or nearly uninterrupted cover, occurs in Arizona generally between 5,000 and 7,000 feet elevation and essentially in the eastern half of the state. Most of it receives 11 to 18 inches of annual precipitation, with extremes of approximately 10 and 21 inches.

San Rafael Valley in Santa Cruz County (Fig. 30), parts of Sulfur Springs Valley in Cochise County, and isolated Chino Valley in Yavapai County are examples of plains grassland habitats carpeted with species of grama grass (*Bouteloua*), muhly (*Muhlenbergia*), needlegrass (*Stipa*), dropseed (*Sporobolus*), sprangletop (*Leptochloa*), and others.

Fig. 26. Desert-grassland isolated in the Sonoran Desert. Lower Sonoran. Tobosa grass (*Hilaria mutica*), foot-high, in the swale of a periodically flooded desert valley at Ventana Ranch, 2,100 ft., roughly midway between Ajo and Sells, in *southwestern* Arizona. A few large palmilla (*Yucca elata*) remain in parts of this tobosa swale which is surrounded by desertscrub of creosotebush, paloverde, sahuaro, *et al.* on higher (coarser-soiled and better-drained) ground. This tobosa landscape, on bottomland, is a principal one in the desert-grassland (Upper Sonoran) in *southeastern* Arizona; e.g., in Cochise County, where such swales are surrounded by other desert-grassland habitats (as in Figures 23 and 24). The dark vegetation line across the "skyline" is a distant mesquite bosque.

In northeastern Arizona, particularly in Navajo and Apache Counties, the formerly extensive plains grassland is now much reduced. There the native grasses are mostly species of grama (*Bouteloua*), fescue (*Festuca*), dropseed (*Sporobolus*), wheatgrass (*Agropyron*), muhly (*Muhlenbergia*), and brome (*Bromus*); galleta (*Hilaria jamesi*) is also one of the characteristic species and its counterpart in the southeastern desert-grassland is tobosa (*Hilaria mutica*). Plains grassland is a semi-arid grassland habitat which occasionally extends upward into the lower portion of the Transition Zone (ponderosa pine forest) in northern Arizona, and is also often in various mixtures with juniper-pinyon woodland and sagebrush.

Fig. 27. Desert-grassland, Upper Sonoran, Empire Mountains, 12 miles southeast of Vail, Pima County, *ca.* 4100 ft., looking southeast across Davidson Canyon toward highest Empire Peak, March 21, 1915. In immediate foreground is a "Rogues gallery of noxious invaders" of the grassland: from left to right, Thornber yucca (*Yucca baccata thornberi*), cane cholla (*Opuntia spinosior*), velvet mesquite (*Prosopis juliflora vetulina*), prickly pear (*Opuntia engelmanni*), Parry agave (*Agave parryi*), ocotillo (*Fouqueria splendens*), and white-thorn (*Acacia constricta*). In background are larger mesquites (deciduous), Mexican crucillos (*Condalia spathulata*, evergreen), and one-seed juniper (*Juniperus monosperma*, evergreen). Today at this locality, which can be viewed from State Highway 83 between Vail and Sonoita, there is an obvious increase of the dry-tropic scrub invaders on the bajada, in the draws and on the lower slopes, unquestionably at the expense of the perennial grasses of forage value. Photo by Forrest Shreve.

Fig. 28. Desert-grassland. Upper Sonoran. A desert-like landscape on a rocky south-facing slope within the desert-grassland, 17 miles N of Sonoita, *ca.* 4,000 ft., in Pima County. Grasses are virtually lacking and ocotillo and prickly-pear are conspicuous, as is commonly the case on such sites today. Other shrubs and cacti here are golden-flowered agave, sotol, mesquite, Mexican crucillo, white-thorn, chollas and barrel cactus.

Fig. 29. Desert-grassland. Upper Sonoran. Sand sage (*Artemisia filifolia*) within the desert-grassland on deep, fine sand along the eastern edge of Willcox Playa, *ca.* 4,200 ft., Cochise County; looking eastward toward the Dos Cabezas Mountains. A few large mesquites and a tall palmilla are conspicuous.

Fig. 30. Plains Grassland. Upper Sonoran. San Rafael Valley, 5,000 ft., east of Lochiel, Santa Cruz County. Looking southwestward into Sonora from south side of Huachuca Mountains near mouth of Parker Canyon.

Mountain Grassland. Mountain grassland occurs in relatively small areas which are natural openings in coniferous forest. These are found from the ponderosa pine forest well into spruce-alpine fir forest. Greatest development is reached in the White Mountains, *e.g.*, the "prairies" of the Apache National Forest, and on the Kaibab Plateau (Fig. 31). It is also fairly well represented in some of the higher isolated mountain ranges in southern Arizona such as the Pinaleno Mountains and Chiricahua Mountains. The forest edge is well marked where the two habitats come together and produce the forest "edge effect." The soils have relatively high rates of moisture evaporation, and obviously have physical properties unsuitable for tree growth.

The characteristic grasses include mountain timothy, Arizona fescue, mountain muhly, pine dropseed, black dropseed, needlegrass, mountain brome, Arizona wheatgrass, and the introduced Kentucky bluegrass. Herbs are common and during the summer beautiful fields of flowers may rise above the green grass carpet. This is particularly true in the Transition and Canadian Life-zones. At higher elevations, where these "mountain meadows" border spruce-fir forest, the wetter (lower) sites are often actually dominated by a considerable number of mostly low-growing herbs instead of by the grasses and sedges; the somewhat drier (higher) drainage sites are those dominated by the grasses and grass-like species.

Chaparral

Interior chaparral occurs in Arizona[17] in the central part of the state, usually between 4,000 and 6,000 feet elevation, from the foothills below the Mogollon Rim to somewhat south of the Gila River, and from the eastern border of the state westward, in progressively smaller areas, into Mohave County where it is fairly well developed as far west as the Hualapai Mountains. Small stands may occur to as low as 3,500 feet and to as high as 7,000 feet, and precipitation ranges from approximately 13 to 23 inches annually. The physiognomy of chaparral is that of dense shrubby growth, usually closed or not widely open, of fairly uniform height between 3 and 6 or 7 feet, broken by an occasional taller shrub or short tree. In Arizona, fairly uniform tall stands of curl-leaf mountain mahogany (*Cercocarpus ledifolius*) may reach over 10 feet in height.

The dominant plants are generally tough-leaved evergreen shrubs. Scrub oak (*Quercus turbinella*) is by far the most common dominant and it may account for over 90 per cent of the stand in many areas (Fig. 32); it is more than likely that it is conspecific with the common scrub oak (*Q. dumosa*) in the California coastal chaparral.

The following twenty-odd species are mostly evergreen and, with scrub oak, are among the most conspicuous or common shrubs in the Arizona chaparral: manzanita (*Arctostaphylos pungens; A. pringlei*),

[17] There is considerable confusion as to what constitutes chaparral in Arizona.

Fig. 31. Mountain Grassland. Hudsonian Life-zone. Looking southward in V T Park, Kaibab Plateau, north of Grand Canyon, 9,000 ft. The bordering coniferous forest is dominated primarily by blue spruce (*Picea pungens*).

Fig. 32. Chaparral. Upper Sonoran. Essentially pure stand of scrub oak (*Quercus turbinella*) on lower slopes of the Prescott Mountains, midway between Iron Springs and Skull Valley, *ca.* 5,000 ft., Yavapai County. There is slight variation in stature in this chaparral which stands between knee-height and hip-height.

sugar sumac (*Rhus ovata*), scarlet sumac (*Rhus glabra*), squawbush (*Rhus trilobata*), mountain-mahogany (*Cercocarpus breviflorus* and *C. betuloides*), buckbrush (*Ceanothus greggi*), deerbrush (*Ceanothus integerrimus*), buckthorn (*Rhamnus crocea, R. californica, R. betulaefolia*), silk-tassel (*Garrya wrighti, G. flavescens*), Apache plume (*Fallugia paradoxa*), brickellbush (*Brickellia californica*), red mahonia (*Berberis haematocarpa*), wait-a-minute bush (*Mimosa biuncifera*), mountain-balm (*Eriodictyon angustifolium*), cliffrose (*Cowania mexicana*), poison-oak (*Rhus diversiloba*) and turpentine bush (*Aplopappus laricifolius*). California fremontia (*Fremontodendron californicum*) and mock-locust (*Amorpha californica*) are shrubs of the chaparral in California and Baja California which occur also in the interior chaparral of Arizona as local species usually occurring in canyons. Sugar sumac, buckthorn, mountain-mahogany (*C. betuloides*), and brickellbush, as listed above, are among the several other disjunctive species between California coastal chaparral and Arizona interior chaparral today separated by desert.

There is no species of grass peculiar to the chaparral formation in Arizona. Grasses may be scarce in closed chaparral, yet may be abundant in open chaparral and especially so following burns. Blue, black, and sideoats grama, plains lovegrass, wolftail, cane beardgrass, red-brome, desert fluff grass, bush muhly and red three-awn are among the commonest species found either in parts of, or throughout, the chaparral.

Chaparral occurs widely in coastal California and adjacent Baja California (coastal chaparral), and under a similar effective climate in the Mediterranean region where it is called *macchie* and *garique*. Our term chaparral evolved from the Basque word *chabarra* and the later Spanish word *chaparro*, both for dwarf evergreen oaks. In the New World, the place name suffix *-al*, meaning "place of," was added and our present word chaparral resulted. It is a useful designation for a distinctive climax vegetation and biotic community occurring under a more or less distinctive climate (Cronemiller, 1942).

Evergreen Woodland

Southern Arizona. In southern Arizona the woodland trees are either mostly or wholly evergreen oaks (*Quercus*, Fig. 33). It may often be an *oak woodland* comprised primarily of Emory oak (*Quercus emoryi*, the most common species, Fig. 34). In addition, there may be Arizona oak (*Q. arizonica*), and Mexican blue oak (*Q. oblongifolia*), with alligator juniper (*Juniperus deppeana*) occasional to abundant, and one-seed juniper (*J. monosperma*) and Mexican pinyon (*Pinus cembroides*) of sporadic occurrence (encinal, Fig. 35). It is usually an open (and often very open) woodland, with numerous associated species of grasses, drytropic shrubs, succulents, and some cacti more or less prevalent throughout.

Fig. 33. Oak Woodland. Upper Sonoran. Large evergreen oaks in foreground, "Stone Cabin Canyon" (Florida Canyon) on north side Santa Rita Mountains, at *ca.* 4100 ft. In upper half of view is typical gradual transition (continuum) from (1) desert-grassland at extreme right center, through (2) open oak-grass savanna (open encinal) in center, to (3) oak woodland (dense encinal) above, on the center peaks and on slopes at extreme left center, with (4) coniferous forest on high slopes and ridges at top left. Photo by Forrest Shreve, 1911. Investigation of this area at present reveals three essential points with regard to change in the general vegetation: (1) dry-tropic shrub and tree density has increased in the desert-grassland area, with reduction of perennial grasses; (2) evergreen oaks (Emory oak, Mexican blue oak) have recently died in the marginal area of oak-grass savanna; and (3) mortality of evergreen oaks within the oak woodland community remains unchanged; there is very little deterioration within, and no decimation of, the oak woodland itself.

Fig. 34. Oak Woodland. Upper Sonoran. An extensive stand of evergreen oaks which is nearly a pure stand of Emory oak (*Quercus emoryi*), *ca.* 5,500 ft. and below, from south side Huachuca Mountains (between Moctezuma Pass and Sunnyside) southward into Sonora. Arizona oak (*Q. arizonica*) and Mexican blue oak (*Q. oblongifolia*) are also present.

Mexican oak-pine woodland lies between the oak woodland (or encinal)[18] below and the ponderosa pine forest (Transition Life-zone) above (Fig. 36). It is characterized in part both floristically and vegetatively by the presence of two large mid-elevation conifers, Chihuahua pine (*Pinus leiophylla*) and Apache pine (*Pinus engelmanni*); these species, with Mexican pinyon and alligator juniper, are variously intermingled with several species of evergreen oaks — principally silverleaf oak (*Q. hypoleucoides*), Arizona oak (*Q. arizonica*), and Emory oak (*Q. emoryi*).

All of these evergreen woodland types (oak, encinal, pine-oak) are largely dominated by species of evergreen oaks and are primarily situated in the southeastern quarter of the state south of the Gila River, where they occur on hills and mountain slopes between 4,000 and 6,500 feet (occasionally slightly higher) and reach their greatest development

---

[18] The term *encinal* refers to a primary Sierra Madrean type of woodland in the sub-Mogollon Southwest. It is dominated by oaks, junipers, and pinyons in more or less equal or subequal abundance, in which the former two or the latter two trees may predominate locally in association with a number of chaparral shrubs (Shreve, 1915; Marshall, 1957; Lowe, 1961). These papers and those of Gentry (1942) and Leopold (1950) refer also to oak-pine woodland. See in particular the excellent work of Marshall (1957).

Fig. 35. Encinal. Upper Sonoran. A rocky habitat with shallow granitic soil, south side Santa Catalina Mountains, 5,800 ft. This is a mixed evergreen woodland dominated by pinyon, oaks, and juniper. The principal tree in foreground is Mexican pinyon (*Pinus cembroides*), often more prevalent than juniper at 5,500-6,000 ft. and above. The oaks are Emory oak and Arizona oak and the juniper is alligator juniper (*Juniperus deppeana pachyphlaea*). Shreve's term "encinal" is appropriate; it refers to Sierra Madrean evergreen woodlands that are partly or wholly dominated by evergreen oaks. This woodland has also been called "oak-juniper type"; one-seed juniper (*J. monosperma*) is the second species locally associated with evergreen oaks between 4,500 and 5,500 ft., as in the Santa Rita Mountains.

on the foothills of the larger mountains such as the Pinals, Pinalenos, Galiuros, Santa Catalinas, Baboquivaris, Santa Ritas, Huachucas, and Chiricahuas.

The shrubs which center their distributions in these oak type woodlands usually range upward into the pine forest or downward into desert grassland or both. Characteristic species are velvet-pod mimosa (*Mimosa dysocarpa*), woodland sumac (*Rhus choriophylla*), algerita (*Berberis haematocarpa*), mountain yucca (*Yucca schotti*), golden-flowered agave (*Agave palmeri*), and Parry agave (*Agave parryi*). Buckbrush (*Ceanothus fendleri*) and locust (*Robinia neomexicana*) are species occurring in the ponderosa pine forest which reach their lower elevational limits in woodland. A number of primarily chaparral shrubs may occur in parts of the oak type woodlands; buckthorn (*Rhamnus*), manzanita (*Arctostaphylos*), mountain mahogany (*Cerocarpus*), squawbush (*Rhus*), poisonoak (*Rhus*), and silktassel (*Garrya*) are representative. Some of these grow primarily under the direct canopy of the oak tree (*e.g.*, squawbush, *Rhus trilobata*), as does the native canyon grape (*Vitis arizonica*), while others grow primarily in open areas between the trees (*e.g.*, Mexican manzanita, *Arctostaphylos pungens*).

Other dry-tropic shrubs and succulents occurring to varying degrees in these interior Southwest woodlands, and which are features of the woodland at its desert or grassland edge (as is the tree *Vauquelinia californica,* rosewood), include coral-bean (*Erythrina flabeliformis*), ocotillo (*Fouquieria splendens*), mesquite (*Prosopis juliflora,* shrubform or tree), wait-a-minute bush (*Mimosa biuncifera*), Wislizenius dalea (*Dalea wislizeni*), feather dalea (*Dalea formosa*), turpentine bush (*Aplopappus laricifolius*), sotol (*Dasylirion wheeleri*), Thornber yucca (*Yucca baccata thornberi*), and palmilla (*Yucca elata*).

The cacti here include woodland and forest subspecies of the pincushion (*Mammillaria vivipara*), cream cactus (*Mammillaria heyderi*), and hedgehog (*Echinocereus troglichidiatus*), as well as species from the desert and grassland such as sahuaro (*Cereus giganteus*), tree cholla (*Opuntia versicolor*), and Wislizenius barrel cactus (*Echinocactus wislizeni*) which are occasional along the lower edge of the woodland; prickly pear cactus is often present. Conspicuous perennial herbs include species of beardtongue (*Penstemon*), verbena (*Verbena*), globemallow (*Sphaeralcea*), lupine (*Lupinus*), mint (*Salvia*), mariposa (*Calochortus*), *et. al.*

A few grasses, such as bullgrass (*Muhlenbergia emersleyi*), little bluestem (*Andropogon scoparius*), and wooly bunchgrass (*Elyonurus barbiculmus*), center in the southerly woodland habitats. Others, such as wolftail (*Lycurus phleoides*) and plains lovegrass (*Eragrostis intermedia*), occur principally in woodland, chaparral, and in the upper parts of desert-grassland. Most of the grasses in the southwestern woodlands, however, are species primarily of the grassland or, less commonly, of the forest;

Fig. 36. Oak-pine Woodland. Upper Sonoran. South side Santa Catalina Mountains, 6,000 ft. The tall pine at left-center is Chihuahua pine (*Pinus leiophylla chihuahuana*). The silhouetted tree with rounded crown at right-center is alligator juniper. The oaks are primarily silverleaf oak (*Quercus hypoleucoides*). A few ponderosa pines (upper right) occur at this elevation which is near the lower edge of the forest. Photo by Forrest Shreve.

blue grama (*Bouteloua gracilis*) is the most common grass throughout the evergreen woodland formation in Arizona.

Small stands of Arizona cypress (*Cupressus arizonica*) occur sporadically and primarily at mid-elevations in the evergreen woodland of mountains in southern Arizona, between 3,500 and 7,200 feet elevation. These relictual (post-climax) pockets, which often contain cypress trees up to 70 feet in height, are all restricted to north-facing slopes and/or canyon bottoms where soil moisture is relatively high and the temperatures in both summer and winter are moderate (Fig. 37). The subspecies *Cupressus arizonica arizonica* occurs from northern Mexico to mountains in southern Arizona which are mostly south of the Gila River (in Pima, Cochise, Graham, and Greenlee Counties). The contiguous subspecies *C. a. glabra* occurs between the Gila River and the Mogollon Rim (Coconino County and the adjacent northern parts of Maricopa and Gila Counties).

Precipitation in woodland habitats in southern Arizona is bi-seasonal and usually somewhat greater in summer than in winter, with annual rainfall usually between 12 and 22 inches; winter precipitation is predominantly in the form of rain rather than snow. The essentially Sierra Madrean characteristic of these woodlands is importantly determined both by the quite moderate minimum winter temperatures which ordinarily prevail, and by the southwestern summer monsoon with its principal rainfall during warm July-August.

Northern Arizona. In northern Arizona where minimum winter temperatures are markedly lower and summer temperatures are also lower, the woodland is composed essentially of junipers and pinyons (Fig. 38, 39). Annual precipitation varies from 12 to 20 inches and in winter it is predominantly in the form of snow. *Juniper-pinyon woodland* covers large areas below ponderosa pine forest on (and near) the Mogollon, Coconino, and Kaibab plateaus, most characteristically between 5,500 and 7,000 feet, and it covers the often flat-topped mesas and plateaus of Navajo and Apache Counties between 5,800 and 7,200 feet elevation. It is also present in several other northern areas including parts of the Strip Country north and west of the Grand Canyon. The woodlands as a whole are among the simplest vegetations in the Southwest, as far as dominant plants are concerned, and juniper-pinyon is perhaps the simplest of the woodlands. Below 6,500-6,800 feet the junipers are more abundant than the pinyons and may occur in pure stands, often as a juniper-grass "savanna." More or less dense stands of juniper (juniper woodland), and open stands of juniper (juniper grassland), occur as minor associations in both northern and southern Arizona (*e.g.*, Pima and Santa Cruz Counties) below 6,500 feet. Above this elevation the pinyons reach their greatest size and they also sometimes grow in large pure stands. By and large, in Arizona as elsewhere in the Southwest, junipers are generally more prevalent and more important in making up the juniper-pinyon woodland matrix than is pinyon (Woodin and Lindsey, 1954).

Colorado pinyon (*Pinus edulis*) is the common and characteristic species of pinyon almost throughout, and Utah juniper (*Juniperus osteosperma*) and one-seed juniper (*J. monosperma*) are the common and widespread junipers. Singleleaf pinyon (*Pinus monophylla*) occurs locally with Utah juniper, mostly in the northwestern corner of the state. Rocky

Fig. 37. Arizona Cypress in the evergreen woodland. Upper Sonoran. An essentially pure stand. Bear Canyon, 5,450 ft., north-facing slope in Santa Catalina Mountains. This beautiful tree (*Cypressus arizonica*) occurs in relict stands (postclimax) now restricted to north-facing slopes and riparian habitats at mid-elevations in Sub-Mogollon evergreen woodlands. Greater stature (to 90 feet) is obtained by those growing in canyon bottoms.

Fig. 38. Juniper-pinyon Woodland. Upper Sonoran. Utah juniper (*J. osteosperma*) and pinyon (*Pinus edulis*) in Skull Valley, Yavapai County. A common feature of juniper-pinyon woodland is simple structural and floristic composition, usually with a "ground cover" essentially of grass as shown here, or a simple shrub understory such as sagebrush (*Artemisia*). Photo by Robert R. Humphrey.

Mountain juniper (*J. scopulorum*) is scattered primarily in the northeast. Alligator juniper[19] and Mexican pinyon, which are the species commonly occurring with evergreen oaks in the mountains of the south, are absent in the north; alligator juniper still occurs today in the central part of the state. It is well known that juniper, as well as mesquite and other southwestern trees, have widely extended onto some of Arizona's former range grasslands (Miller, 1921; Pearson, 1931; Parker, 1945; Arnold and Schroeder, 1955; Humphrey, 1962; and others).

Grasses are (or were) more or less abundant throughout juniper-pinyon woodland. The predominant species include blue grama (one of

[19] In addition to alligator juniper (*J. deppeana pachyphlea*), one-seed juniper (*J. monosperma*) of the north is occasionally present in southern Arizona, in the edge of the desert-grassland and lower edge of the woodland (not in mountains) in Cochise County and western Pima and Santa Cruz Counties.

Fig. 39. Juniper-pinyon Woodland and Great Basin Desertscrub. Upper Sonoran. Looking northeast from west of Cameron, Coconino County, across Little Colorado River Gorge, and across the "Painted Desert" to its bordering plateaus ("the Hopi Mesas"). The "Painted Desert," which lies essentially below 5,000 feet elevation, comprises a relatively small area within the Great Basin Desert. Elevation of juniper-pinyon woodland in foreground approximately 6,000 ft. Photo by E. Tad Nichols.

the most important), sideoats grama, black grama, Arizona fescue, pinyon ricegrass, junegrass, indian ricegrass, needlegrass, sand dropseed, squirreltail, and ring-grass. Commonly a grass such as blue gamma (*Bouteloua gracilis*) or Arizona fescue (*Festuca arizonica*) is the most abundant herbaceous plant present in the stand. The more conspicuous herbs include species of globemallow, beardtongue, mariposa, paintbrush, *et al.*

While the understory shrubs are varied and often numerous (or, indeed, quite sparse and scattered), the dry-tropic species that are so characteristic of the southern oak type woodlands are either entirely absent or weakly represented in the northern woodlands of the state. Characteristic understory shrubs in the juniper-pinyon woodland of central and northern Arizona include cliffrose (*Cowania mexicana*), big sagebrush (*Artemisia tridentata*), serviceberry (*Amelanchier alnifolia*), rabbitbrush (*Chrysothamnus nauseosus, C. depressus*), fernbush (*Chamaebatiaria millifolium*), Navajo ephedra (*Ephedra viridis*), Fremont barberry (*Berberis fremonti*), Apache-plume (*Fallugia paradoxa*), antelope-brush (*Purshia tridentata*), black sage (*Artemisia nova*), banana yucca (*Yucca baccata*), Whipple cholla (*Opuntia whipplei*), beavertail (*O. basilaris*), fragile cholla (*O. fragilis*), prickly pear (*O. polyacantha*), and red hedgehog cactus (*Echinocereus triglochidiatus melanacanthus*).

One or more of the distinctly Great Basin desert species such as big sagebrush (*Artemisia tridentata*), shadscale (*Atriplex confertifolia*), blackbrush (*Coleogyne ramosissima*), and winterfat (*Eurotia lanata*), occur in the lower portion of the woodland; big sagebrush, moreover, occurs throughout approximately the elevational range of this woodland in Arizona. Chaparral forms such as silktassel (*Garrya wrighti*), mountain-mahogany (*Cerocarpus intricatus, C. ledifolius*), scrub oak (*Quercus turbinella*) and Gambel oak (*Q. gambeli,* in thicket shrubform) are commonly present, and species of the ponderosa pine forest such as buckbrush (*Ceanothus fendleri*) and pine cactus (*Opuntia erinacea xanthostema*) occur in the upper part of the woodland.

Deciduous (Riparian) Woodland

The deciduous woodlands are *riparian woodlands,* i.e., they occur along streams, rivers, floodplains and the like. They are comprised mostly of broadleaf trees which are winter deciduous, such as cottonwood, willow, and walnut (Fig. 40, 41, 65, 66). The species and/or genera are several friends in "the woods at home" of the eastern United States. Both their distinctive life-form and their riparian habitat distinguish these woodlands immediately from the evergreen western woodlands. The associated animals are conspicuously different in the two biotic communities, although the one (riparian) may course, finger-like, through the other. Zonally the habitat extends from the Lower Sonoran into Canadian, rarely into the Hudsonian zone.

The trees are often large, some species reaching heights of 50 to 100 feet (*e.g.,* cottonwood, sycamore, and alder). Thus the woodland may be a fairly high-canopied gallery association. Excluded from this formation are low shrubby stream thickets without trees, mountain meadows, and aspen (or maple, or other tree) stands or thickets which occur on slopes without immediate stream development.

Fig. 40. Riparian Woodland. Upper Sonoran. The cottonwood-willow gallery association of broadleaf, winter deciduous trees on Sonoita Creek, 4,000 ft., just west of Patagonia, Santa Cruz County. Mesquites may form an intermittent lower story on the stream bank beneath the broadleaf canopy (center-right), as well as forming extensive "mesquite bosques" on the adjacent flood plains. The surrounding terrain is desert-grassland and oak woodland. The stream is permanent and supports five species of native fishes and some introduced panfishes. Photo by E. Tad Nichols.

This is the least xeric of the woodlands, and one of the most mesic of all the strictly terrestrial ecologic formations in the Southwest. The habitat is in and along the channels, their margins, and/or floodplains of the larger and/or better watered drainageways. In southern Arizona most of these drainageways are the permanent and intermittent streamways of the conifer-clad mountains and their arroyo extensions into rivers and river beds which may still receive a sufficient flow above or below ground surface.

The species composition changes with elevation. Riparian woodland distributed along major drainageways throughout the evergreen woodlands and much of the conifer forests is characterized by such deciduous broadleaf trees as Texas mulberry (*Morus microphylla*), Arizona alder (*Alnus oblongifolia*), narrowleaf cottonwood (*Populus angustifolia*), southwestern chokecherry (*Prunus serotina virens*), boxelder (*Acer negundo*), Rocky Mountain maple (*Acer glabrum*), and Scouler willow (*Salix scouleriana*).

Within the broadleaf riparian woodland, isolated oak trees (*e.g.*, *Q. emoryi, Q. arizonica, Q. oblongifolia, Q. hypoleucoides*) occasionally finger down into the desert as far as 1,000 feet below the oak zone, and scrub oak to an elevation of 2,500 feet or lower (*e.g., Q. turbinella ajoensis*). Mesquite (*Prosopis juliflora*), catclaw (*Acacia greggi*) and others, often form a distinctive microphyllous border association on adjacent floodplains.

In the center of the Sub-Mogollon region, the riparian "big-five" are cottonwood (*Populus fremonti*), willow (*Salix bonplandiana* and others), sycamore (*Platanus wrighti*), ash (*Fraxinus pennsylvanica velutina,* and others), and walnut (*Juglans major*). Often three or four of these species may occur together, and occasionally all five. All of them are large, winter-deciduous broadleaf trees of genera and families different from those of the immediately bordering non-riparian climax desert or grassland communities. (Fig. 41).

This broadleaf association forms the conspicuous and dominant plant component of a biotic community which includes characteristic species of aquatic, and semi-aquatic, and terrestrial animals. Examples of these are the summer tanager (*Piranga rubra*), Bullock oriole (*Icterus bullocki*), yellow warbler (*Dendroica petechia*), Sonoran mud turtle (*Kinosternon sonoriense*), black-necked garter snake (*Thamnophis cyrtopsis*), leopard frog (*Rana pipiens*), canyon treefrog (*Hyla arenicolor*), longfin dace (*Agosia chrysogaster*), other vertebrates, and a larger number of invertebrates.

A riparian association of any kind is one which occurs in or adjacent to drainageways and/or their floodplains and which is further characterized by species and/or life-forms different from that of the immediately surrounding non-riparian climax. The southwestern riparian woodland formation is characterized by a complex of trees, and their plant and animal associates, restricted to the major drainageways that transgress the landscape of desert upward into forest. It is incorrect to regard this biotic formation as merely a temporary unstable, seral community. It is an evolutionary entity with an enduring stability equivalent to that of the landscape drainageways which form its physical habitat. That is, it is a distinctive climax biotic community. Moreover, it is, as are all ecologic formations and their subdivisions, locally subject to, and often dissolved by, the vicissitudes of human occupation. In Arizona, the riparian woodlands have been rapidly dwindling just as the water table has been rapidly lowering. And its trees are now the native phraeatophytes of the water-users.

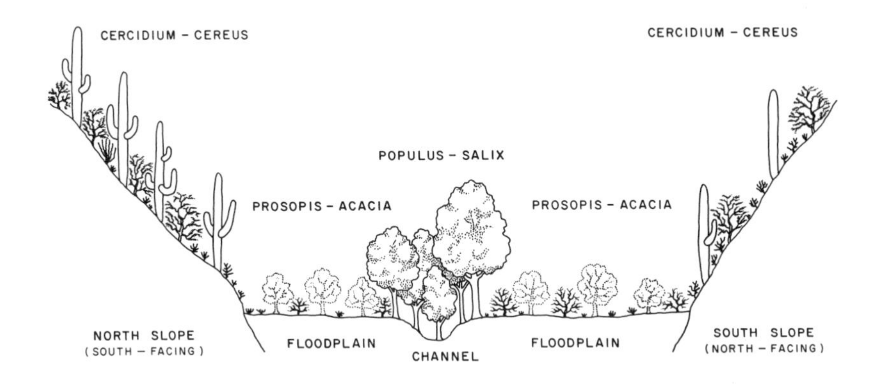

Fig. 41. Diagrammatic profile of a foothill canyon in the Sonoran Desert, as in the Galiuro and Santa Catalina Mountains at approximately 3,000 feet elevation. Streamflow is either permanent or semi-permanent and in either case periodically torrential. The riparian woodland of cottonwood, willow, ash, sycamore, and walnut (cottonwood-willow gallery association) is situated in and at the immediate edge of the channel. This broadleaf deciduous woodland is immediately bordered by a mesquite dominated floodplain characterized by deciduous microphylls (*Prosopis-Acacia* floodplain community). The non-riparian climax on the adjoining slopes is paloverde desert-scrub (*Cercidium-Cereus* upland community) dominated primarily by aphyllous and microphyllous deciduous and often spinose life-forms of trees, shrubs, and cacti.

Transition Life-zone

The Transition Zone is equivalent to *pine forest,* and specifically to *ponderosa pine forest* (western yellow pine forest). Thus the entire zone in Arizona, and often elsewhere, is the *association* and also frequently the *consociation* (pure stand) dominated by the single species *Pinus ponderosa.* Mature trees reach large size (to 125 feet), are commonly 200 to 400 years old, and may exceed 500 years in age.

Elevation of the forest is ordinarily between 6,000 and 9,000 feet; 6,000-7,000 feet at the lower extreme, and 8,500-9,000 feet at the higher, varying according to slope exposure. Pure stands of ponderosa, however, occur most commonly between 7,000 and 8,000 feet. The lower limit of the ponderosa pine forest is the lower elevational limit of the entire coniferous forest formation within Arizona. This lower ecologic limit is controlled by several factors of which critical low level of plant-available soil moisture is the primary direct factor (Shreve, 1915; Sampson, 1918; Pearson, 1920, 1931; Daubenmire, 1943a). Precipitation within this zone is approximately 18 to 26 inches annually; rarely lower.

Fig. 42. Ponderosa Pine Forest. Transition. View from Sunset Crater toward east face of San Francisco Mountain (San Francisco Peaks) across part of volcanic field of same name, 7,000 ft. In the foreground is the cinder-covered Bonita lava flow which emerged from the base of Sunset Crater and which has remained largely unforested. It is early summer and patches of snow remain on the high peaks bordering the Inner Basin of this old volcano which once stood about 3,000 feet higher than at present—12,670 feet on Humphreys Peak (right center). Photo by E. Tad Nichols.

This is a major forest type that covers much of the Kaibab Plateau and Mogollon Mesa, and other areas on the extensive Colorado Plateau (Fig. 42, 43); Rocky Mountain ponderosa pine (*Pinus ponderosa scopulorum* Engelm.) is the subspecies in this area. Transition is also the principal (largest area) forest life-zone on the higher conifer-clad mountains of the state, including several of the desert-border ("desert") mountains located south of the Salt River (Fig. 44, 45). Arizona ponderosa pine (*Pinus ponderosa arizonica* Engelm.) is the primary subspecies (typically 5-needled) in the southern area (primarily in Pima, Santa Cruz, and Cochise counties).[20] In addition to the different tree subspecies which occur, there are several

[20] Subspecies (subsp.) of plants may still be referred to as varieties (var.) in some references. Some mountains in southern Arizona, such as the Santa Catalinas near Tucson, have two subspecies of ponderosa pine: one (*P. p. scopulorum*) at the higher elevations mostly above 8,000 feet elevation on south-facing slopes, and the other (*P. p. arizonica*) at lower elevations mostly between 6,000 and 8,000 feet.

Fig. 43. Ponderosa Pine Forest. Transition. Navajo County. This is a landscape characteristic of the ponderosa forest across *central and northern* Arizona on flat and gently rolling plateau country; grass is the principal "ground cover." The relatively open, park-like stand across more or less level expanses is a feature almost entirely lacking in the forested mountains of *southern* Arizona where, in sharp contrast, the gradients are steep, soils more shallow, and shrubs more common. Photo by Robert R. Humphrey.

notable differences between the Transition Zone habitats of the southern mountains south of the Salt River and those to the north, on the Colorado Plateau.

Southern Arizona. The southern conifer-clad mountains, such as the Pinals, Gilas, Pinalenos, Galiuros, Santa Catalinas, Santa Ritas, Huachucas, and Chiricahuas, are isolated ranges[21] with relatively steep topography when compared to the extensive mesas and plateaus to the north. There are few level areas with parks or mountain meadows. Flood plains are small and not well developed, for typically the streams are narrowly confined and their gradients are steep.

These rugged ranges are populated with many species of plants and animals that have their areas of principal distribution in Mexico and

[21] "Sky islands" (Heald, 1951), seems to me quite apt. They are often called "desert mountains" or "desert mountain islands" which are inappropriate, for they are not surrounded by desert. They are bordered on one or more sides by grassland as well as by desertscrub on the other sides.

Fig. 44. Ponderosa Pine Forest. Transition. Ponderosa pine (*Pinus ponderosa arizonica*) in the *lower* part of the forest, 7,000 ft., Santa Catalina Mountains, Pima County. Silverleaf oak (*Quercus hypoleucoides*) is the characteristic broadleaf tree. Madrone (*Arbutus arizonica*) and netleaf oak (*Q. rugosa*) are also present along with alligator juniper and Chihuahua pine, all evergreen. Photo by Forrest Shreve.

reach their northern limits in Arizona below and south of the Mogollon Rim. At lower elevations, silverleaf oak (*Quercus hypoleucoides*), netleaf oak (*Quercus reticulata*), and madroño (*Arbutus arizonica*) are usually the commonest broadleaf evergreen trees. At higher elevations, and in cooler and more mesic situations, Gambel oak (*Quercus gambeli*), bigtooth maple (*Acer grandidentatum*), aspen (*Populus tremuloides*), alder (*Alnus oblongifolia*), and mulberry (*Morus microphylla*) are common broadleaf deciduous trees. Similarly, among the coniferous associates, Chihuahua pines are occasionally scattered among the ponderosa pines at the lower zone limit and alligator juniper is scattered throughout the lower, more xeric parts of the zone, while Douglas fir (*Pseudotsuga menziesi*) and white pine (*Pinus ayacahuite*) become fairly frequent in the upper and more mesic parts. In these ranges, ponderosa parklands with well-developed grass areas and/or shrub understories, so typical in the north, are more often virtually non-existent.

Buckbrush (*Ceanothus fendleri*) is the small shrub that is prominent throughout the southern pine forest to about 9,000 feet; it may be the only shrub present on the forest floor, and may form thickets on slopes

Fig. 45. Ponderosa Pine Forest. Transition. Typical dense stand in the *upper* part of the forest, between 8,000 and 9,000 ft., Santa Catalina Mountains. In addition to dominant ponderosa pine (*Pinus ponderosa scopulorum*), white pine (*Pinus ayacahuite*) and Douglas fir (*Pseudotsuga menziesi*) are present. The deciduous Gambel oak (*Quercus gambeli*) is a characteristic broadleaf tree, and evergreen oaks (silverleaf and netleaf) are also present; but neither alligator juniper, pinyon, nor Chihuahua pine reach 8,000 ft. elevation. Photo by Forrest Shreve (1915, plate 35).

where the forest is particularly open. Boxleaf myrtle (*Pachystima myrsinites*) is a low, creeping, inconspicuous undershrub that is more commonly seen in the upper part of the forest. Arizona rose (*Rosa arizonica*) along streams, and New Mexican locust (*Robinia neomexicana*) as a large and conspicuous shrub or small tree occur throughout much of the forest. Between 8,500 and 9,000 feet, snowberry (*Symphoricarpos oreophilus*), ocean spray (*Holodiscus dumosus*), orange gooseberry (*Ribes pinetorum*), and other shrubs which are also common to fir forest and spruce-fir forest, enter the upper portion of ponderosa pine forest.

In the lower part of the forest, shrubs of the woodland or chaparral are often present, such as buckthorn (*Rhamnus crocea, R. betulaefolia, R. californica*), deerbrush (*Ceanothus integerrimus*), manzanita (*Arctostaphylos pungens, A. pringlei*), squawbush (*Rhus trilobata*), silktassel (*Garrya wrighti*), mountain yucca (*Yucca schotti*), and hedgehog cactus (*Echinocereus triglochidiatus*). Scarlet sumac (*Rhus glabra*) occasionally forms thickets in the lower edge of the forest. Also here, two of the prominent oaks of the pine forest (the facultative species *Q. hypoleucoides* and *Q. reticulata*) may be in shrub form rather than tree form, and western chokecherry (*Prunus serotina*) also may be a shrub or small tree.

In many of the genera of perennial herbs common to the southern and northern ponderosa pine forests in Arizona and New Mexico, the species and/or subspecies may be different; for example, as in lupine (*Lupine*), peavine (*Lathyrus*), cinquefoil (*Potentilla*), yarrow (*Achillea*), goldenrod (*Solidago*), paintbrush (*Castilleja*), beardtongue (*Penstemon*), fleabane (*Erigeron*), deervetch (*Lotus*), groundsel (*Senecio*), pinque (*Actinea*), milkvetch (*Astragalus*), violet (*Viola*), and many others. Conversely, many of them which occur in the two forest areas are of the same species; for example, bracken fern (*Pteridium*), Fendler globemallow (*Sphaeralcea fendleri*), beebalm (*Monarda*), flag (*Iris*), mullein (*Verbascum*, introduced), toadflax (*Commandra*), mock-pennyroyal (*Hedeoma*), golden pea (*Thermopsis*), and other species of lupine, yarrow, *et al.*

The most characteristic grasses are the two mountain muhlys (*Muhlenbergia montana, M. virescens*). Other characteristic grasses are pine dropseed (*Blepharoneuron tricholepis*) and pinyon ricegrass (*Piptochaetium fimbriatum*). Most of the characteristic grasses that are present in both the southern Arizona and northern Arizona pine forests are conspecific.

Northern and Central Arizona. In central and northern Arizona north of the Mogollon Rim, the ponderosa forests usually contain fewer trees of other species, but they are rich in shrubs and in the grass carpeting which often extends through park-like landscapes. On the few high mountains which rise abruptly from the Colorado Plateau in northern Arizona, *e.g.*, the San Francisco Mountains and Chuska Mountains, the Transition Zone of ponderosa pine forest is also present on rugged precipitous terrain as it is on the mountains in the Basin and Range Province in southern Arizona. Such is not the usual situation, however, for, as already noted, Transition Zone in northern Arizona is typically distributed over very extensive areas of flat to rolling plateau country.

Gambel oak is the most common tree associated with the pines, and quaking aspen may be scattered or may be in large stands on old burns, usually above 7,500 feet. In the lower part of the zone, below 7,000-7,500 feet, pinyon, juniper, and big sagebrush may be mingled with the pines; Douglas fir is occasional to frequent above about 7,000 feet.

Understory shrubs, in climax stands, may be essentially lacking to fairly common though more or less widely and irregularly spaced. Also, some shrubs (*e.g.*, species of *Artemisia*) may be locally abundant, particularly at lower elevations in the forest as seen on the South Rim of Grand Canyon and in the Navajo Indian Reservation where ponderosa pine occurs. And other shrubs (*Ceanothus, Berberis*) may be more or less uniformly distributed over large areas.

Among the most characteristic shrubs of this northern Arizona forest are buckbrush (*Ceanothus fendleri*), fernbush (*Chamaebatiaria millefolium*), Gambel oak (*Quercus gambeli,* shrubform), fendlerella (*Fendlerella utahensis*), wax currant (*Ribes cereum*), New Mexican locust (*Robinia neomexicana*), Canadian elder (*Sambucus coerulea*), greenleaf manzanita (*Arctostaphylos patula*), Parry rabbitbrush (*Chrysothamnus parryi*), big sagebrush (*Artemisia tridentata*), black sagebrush (*A. nova*), cliffrose (*Cowania mexicana*), Apache-plume (*Fallugia paradoxa*), and mockorange (*Philadelphus microphyllus*). At higher elevations the shrubs of the pine forest are more commonly species shared with the fir forest and spruce-alpine fir forest: ninebark (*Physocarpus monogynus*), raspberry (*Rubus strigosus*), shrubby dwarf juniper (*Juniperus communis montana*), ocean spray (*Holodiscus dumosus glabrescens*), Oregon grape (*Berberis repens*), cliffbush (*Jamesia americana*) and boxleaf myrtle (*Pachystima myrsinites*).

Grasses characteristic of the northern Arizona ponderosa forests include Arizona fescue, mountain muhly, pine dropseed, squirreltail, mountain brome, spike muhly, deer grass, junegrass, and bluestem; blue grama is common in forest openings. Perennial herbs (root perennials) in addition to those already noted in the discussion of the understory plants in the southern Arizona ponderosa forests are meadowrue (*Thalictrum fendleri*), vetch (*Vicia americana*), locoweed (*Oxytropis lamberti*), and sage (*Artemisia ludoviciana*). These are typical pine forest associates more or less throughout the state.

Canadian Life-zone

The Canadian Zone is equivalent to *fir forest*, i.e., *Douglas fir forest*. Elevation is from about 7,500-8,000 to 9,000-9,500 feet (according to slope exposure), and occasionally the zone extends to higher elevations of nearly 10,000 feet. Precipitation is approximately 25 to 30 inches annually.

The fir forest stands are typically of mixed species rather than of a single species as is much of the ponderosa pine forest. Douglas fir (*Pseudotsuga menziesi*) and white fir (*Abies concolor*) are the principal trees (Fig. 46, 47). Douglas fir has the wider tolerance, extends to lower elevation, and predominates on south-facing exposures; white fir may predominate on northerly exposures. Both trees reach large size (about 150 ft. ht.) and stands are commonly 200 to 400 years of age. At the higher elevations, alpine fir (*Abies lasiocarpa*) may be present, and white pines are occasional to frequent; the species in northern Arizona is limber pine (*Pinus flexilis*), and in the southwestern mountains it is southwestern white pine (*Pinus ayacahuite*).

Ponderosa pine may be present, but more or less confined to ridges and southerly exposures except along the lower edge of the fir forest where the two forests tend to merge in the continuum. There are physiographic, ecologic, and taxonomic differences between the fir (Canadian) forests of

Fig. 46. Fir Forest. Canadian. North slopes of Mount Lemmon, Santa Catalina Mountains, 8,700 ft. The trees are white fir (*Abies concolor*), Douglas fir (*Pseudotsuga menziesi*) and white pine (*Pinus ayacahuite*). The slope is gentle at this site but shortly becomes steep as is commonly the case throughout the forest on the mountains of southern Arizona. Photo by Forrest Shreve (1915, plate 1).

northern Arizona and those of southern Arizona which are similar to those already noted for the Transition Zone (ponderosa pine).

Gambel oak (*Q. gambeli*), box elder (*Acer negundo*), water birch (*Betula occidentalis*), and blueberry elder (*Sambucus glauca*), are among the conspicuous broadleaf deciduous trees, as are Rocky Mountain maple at higher elevations and locust at lower elevations. Quaking aspen (*Populus tremuloides*) forms a successional subclimax community (following burns and other disturbances in the forest) that is particularly well developed and conspicuous in this zone prior to re-establishment of the coniferous forest which follows the aspens in time (Fig. 48); this is also true in the Hudsonian Zone and in the upper part of the Transition Zone.

There are few (if any) distinctive understory species either on the floor of the fir forest or on the stream banks and flood plains. In dense stands a considerable duff accumulates on the forest floor similar to that of spruce-fir forests. The shrubs, herbs, and grasses are usually species shared either with the pine forest below (such as bracken fern, vetch, and deers-ears) or spruce-alpine fir forest at higher elevations (such as wolf currant, owlclaws, and red elderberry). In the fir forest, sedges (*Carex*)

Fig. 47. Fir Forest. Canadian. An alluvial flat in fir forest, north slope of Mount Lemmon, Santa Catalina Mountains, 8,600 ft. White fir (*Abies concolor*), ponderosa pine (*Pinus ponderosa*) and quaking aspen (*Populus tremuloides*). Photo by Forrest Shreve.

Fig. 48. Fir Forest. Canadian. Old stand of quaking aspen (*Populus tremuloides*), 9,000 ft., San Francisco Mountain. Note young conifers growing up under canopy of aspens in usual forest succession. Photo by E. Tad Nichols.

become conspicuous and there is no longer evidence of the dry-tropic species of trees, shrubs, succulents, and cacti which enter from below and partly traverse the floor of the yellow pine forest; this is equally true in Sub-Mogollon southern Arizona as well as north of the Mogollon Rim.

The Canadian Zone reaches its greatest development in Arizona on the higher ranges rising from the Colorado Plateau in the central, northern, and eastern parts of the state, and there one of its best developments is in the extensive elevated mass called the White Mountains, southwest of Springerville (Fig. 52). It is occasionally well developed as a forest stand or zone on the higher mountains in the southern part of the state, *e.g.*, on the Santa Catalinas, Santa Ritas, Huachucas, Chiricahuas, and Pinalenos (Fig. 46). Throughout the state, fir forest occurs where it is with few exceptions situated in rugged terrain, and is rarely seen on broad plateaus and flat mesas as is characteristic of ponderosa pine forest in central and northern Arizona.

On the high mountains which "top out" at about 9,000 feet elevation in southern Arizona and in adjacent parts of neighboring states, the fir forest is usually present without spruce-alpine fir forest above it.[22] On such mountains, the fir forest is very distinctive (Fig. 46). It extends downward on the north-facing slopes from the summit to about 7,500 feet elevation, the lower limit varying somewhat with latitude, base-level elevation, mountain mass, etc. The south-facing slopes are covered with ponderosa pine forest. Northward in central and northern Arizona on the Colorado Plateau, however, the fir forest (Canadian Zone) often has a wavering sort of existence intermingled more or less with the spruce-alpine fir forest. As a result of this situation on the Colorado Plateau north of the Mogollon Rim in Arizona and New Mexico, the forest types (Douglas fir-white fir, and spruce-alpine fir) have been considered together as the Northern Mesic Evergreen Forest (Shreve, 1942, Arizona), Petran Subalpine Forest (Castetter, 1956, New Mexico), Boreal Forest, etc.

The environmental tolerances and hence the ecologic and geographic distributions of spruces, firs, Douglas fir, and white pines widely overlap in the Rocky Mountain region. It is not surprising, therefore, that the two mesic forest types in Arizona are not always easily distinguishable in some areas, and that they tend on the whole to contain similar groups of species. Somewhat greater differences are seen in the herb, grass, and sedge components of the two forests.

---

[22] South of the Gila River in Arizona, only two mountain ranges contain spruce (*Picea engelmanni*) in their floras. These are the Pinaleno and the Chiricahua Mountains. The former exceeds and the latter closely approaches 10,000 feet elevation. The spruce stands on the Chiricahuas, in Cochise County, were thought to be the southernmost in North America. Engelmann spruce, however, is reported from the northern part of the Sierra Madre in Chihuahua, Mexico.

Hudsonian Life-zone

The Hudsonian Zone is equivalent to *spruce-alpine fir forest* (Fig. 49, 50). It occurs around and on the summits of the highest ranges, such as the Chiricahua Mountains, Graham Mountains, White Mountains, San Francisco Mountains (summit is alpine tundra), and on the large summit area of the Kaibab Plateau. Elevation is from approximately 8,500 to 9,000 feet at the lower limit to approximately 11,500 feet maximum elevation; precipitation is approximately 30 to 35 inches per annum.

The primary trees are Engelmann spruce, blue spruce, alpine fir (corkbark fir), limber pine, and bristlecone pine (also called foxtail pine). Rocky Mountain maple (*Acer glabrum*), bitter cherry (*Prunus emarginata*), Bebb willow (*Salix bebbiana*), Scouler willow (*Salix scouleriana*), and thin-leaf alder (*Alnus tenufolia*) are among the broadleaf deciduous trees which may be present where the forest is not with completely closed canopy. Aspen occurs as scattered trees, small clumps, or in large and often pure (subclimax) stands.

Engelmann spruce (*Picea engelmanni*) is the principal dominant of the spruce-alpine fir forest, and alpine fir (*Abies lasiocarpa*) is the usually common, characteristic fir tree that is a remarkably constant companion; cornbark fir (*A. l. arizonica*) is a subspecies (variety) of alpine fir occurring in the southern Rocky Mountains, from Colorado to Arizona. Climax spruce-alpine fir forest reaches 80 or more feet in height, varying according to closeness of the trees and other factors. Pure Engelmann spruce stands may well exceed 250 years of age, while alpine fir is a shorter-lived species that tends to die out in old forest stands.

Blue spruce (*Picea glauca*) is a major dominant on the extensive summit area of the Kaibab Plateau, an area which is essentially spruce-fir forest with interspersed mountain grassland (Fig. 49). It is absent from the forests of southern Arizona. Bristlecone pine (*Pinus aristata*) is a conspicuous tree in the upper part of the spruce-fir forest on high San Francisco Mountain, near Flagstaff, and is one of the two gnarled conifers (Engelmann spruce and bristlecone pine) in the prostrate wind-timber (Krummholz) there at timberline (see Arctic-Alpine Zone below).

As noted earlier, the two primary conifers of the Canadian Zone (Douglas fir and white fir) may be present with spruce, and occasionally extensively so, as on Mount Thomas (Baldy Peak) and other areas in the White Mountains, and on the Kaibab Plateau. White pine is also widely distributed in both life-zones and also occurs (but to a much more limited extent) in the upper part of the Transition Zone.

At high elevations on San Francisco Mountain, from above 9,000 to over 10,000 feet elevation, on the steep slopes with well-marked north and south exposures, Engelmann spruce and alpine fir quite typically dominate the forest on northerly exposures all the way to timberline. But on south-facing slopes there is an equally striking forest type above the last Douglas

Fig. 49. Spruce-fir Forest. Hudsonian. Edge of forest on Kaibab Plateau, V T Park, north of Grand Canyon, 9,000 ft. The conifers are blue spruce (*Picea pungens*), the broadleaf trees aspen (*Populus tremuloides*). Blue spruce singularly dominates the forest bordering the mountain grassland, and elsewhere on south-facing slopes. Engelmann spruce (*P. engelmanni*) dominates in other areas on the Kaibab in which alpine fir (*Abies lasiocarpa*) is virtually absent, whereas white fir (*A. concolor*) and Douglas fir (*Pseudotsuga menziesi*) are conspicuously present.

firs that is dominated primarily by limber pine and/or bristlecone pine. This distinctive high pine association may be considered for present purposes a plant community (bristlecone pine-limber pine, *Pinus aristata-Pinus flexilis*) within the spruce-alpine fir biotic community; it usually has been so considered, when not merely ignored. In zonal terminology, bristlecone pine and limber pine form a high pine association in the Hudsonian Life-zone. However, I should point out that habitat-wise, the more open limber pine-bristlecone pine association is not merely intermediate between the climatic and edaphic conditions supporting fir forest and spruce-fir forest. Taken on the whole, such high limber pine forest is, in fact, very distinctive however small its total geographic extent in the Southwest (see Merriam, 1890; Pearson, 1931), and I have listed it in Tables 1 and 2.

Where the forest shade and litter may be reduced or absent — as along streamways, in natural forest openings, on old burns, and in trail cuts — a number of species of shrubs may occur as single plants or in small patches, or more rarely as rather extensive thickets. The characteristic shrubs are species of currant (*Ribes*), blueberry (*Vaccinium oreophyllum*), Oregon-grape (*Berberis repens*), black-fruited honeysuckle (*Lonicera involucrata*), dwarf juniper (*Juniperus communis montana*), red elderberry (*Sambucus racemosa*), and shrubby cinquefoil (*Potentilla fruticosa*). These occur also to either a lesser or greater extent in fir forest (Canadian Life-zone). Toward the upper limit, most of the shrubs have disappeared well before timberline is reached. However, two shrubs extend to and slightly beyond the timberline on San Francisco Mountain: dwarf juniper and gooseberry currant.

Herbs in the lower part of the Hudsonian again are species found also in the Canadian Zone; and the grasses present are mostly those in the Canadian Zone and the Transition Zone. However, the herbs, grasses, and sedges found in the higher parts of the spruce-alpine fir forest between 10,500 and 11,500 feet elevation are often species shared with the alpine tundra community and, in fact, are often primarily alpine tundra species. There are several sedges (*Carex*) and rushes (*Juncus*) and the grasses include mountain timothy (*Phleum alpinum*), red fescue (*Fescue rubra*) and spike trisetum (*Trisetum spicatum*).

Spruce-alpine fir communities are habitats for high elevation primroses (*Primula*), gentians (*Gentiana*), violets (*Viola*), columbines (*Aquilegia*) and other mountain favorites as well as skunk cabbage (*Veratrum*), owlclaws *Helenium*), baneberries (*Actaea*), louseworts (*Pedicularis*) and many others. Most of these herbaceous plants, however, are rare or totally absent in the shade of compact tree stands, as are the shrubs and grasses. In the shade of such climax overstories a heavy duff usually accumulates on the forest floor; sedges, mosses, liverworts and lichens are the characteristic plants and they may be locally abundant.

On San Francisco Mountain where highest Humphreys Peak is 12,670 feet, the Hudsonian Zone timberline varies from about 11,000 feet on northerly slopes to 11,400-11,500 feet on southerly exposures. Slope exposure, toward or away from the sun, is the most important local condition producing the 400-500 foot variation in this upper limit for forest tree growth. Spruce-alpine fir forest occurs with no true alpine

Fig. 50. Spruce-fir Forest. Hudsonian. Rockslide at 11,000 ft. on old Weatherford Road, north slope Fremont Peak, San Francisco Mountain. The trees are Englemann spruce and young ones have established on the old road cut without succession preceded by aspen or other trees. Quaking aspen does not reach this elevation on north-facing slopes, although it occasionally does on south-facing slopes. Photo by H. K. Gloyd.

timberline (Krummholz) at the summit area of Baldy Peak in the White Mountains, although, as noted, conditions there approach timberline.

Arctic-Alpine Life-zone

The Arctic-Alpine Zone is represented in Arizona by a summit area of alpine tundra, an "arctic" type vegetation, isolated on the top of San Francisco Mountain (to 12,670 feet) above its timberline (Fig. 51). This is the only true Arctic-Alpine in Arizona, although tree stature beginning to approach timberline conditions can be seen in the White Mountains (Arizona's second highest) on Baldy Peak (11,470 feet) which is forested to the summit by conifers, and on Mount Graham (10,713 feet) in the Pinaleno (Graham) Mountains south of the Gila River in Graham County.

Geography and geology of the San Francisco Mountain region are given by Robinson (1913), the alpine tundra climate by Merriam (1890), by Coville and MacDougal (1903), and by Pearson (1920, 1931), and the alpine tundra flora and vegetation by Merriam (1890) and by Little (1941).[23] The important paper by Elbert Little (1941) is the principal modern work on the alpine flora and vegetation of San Francisco Mountain.

---

[23] See also Hoffman (1877), Britton (1889), Rusby (1889), Harshberger (1911), Rydberg (1914), Holm (1927), Hesse, Allee and Schmidt (1937), Shreve (1942a), Daubenmire (1954), Castetter (1956), Billings and Bliss (1959), Fosberg (1959), Martin (1959), Webster (1961), and Beaman (1962).

Fig. 51. Alpine Tundra. Arctic-Alpine Life-zone. Humphreys Peak (12,670 ft.), San Francisco Mountain. Looking northwest across Inner Basin toward alpine tundra on Humphreys Peak, highest point in Arizona, from spruce-alpine fir forest on saddle of old Weatherford Road, *ca.* 11,000 ft., between Fremont Peak and Doyle Peak. Note the few lighter patches of quaking aspen among the conifers (Hudsonian Life-zone) on southeast-facing slope coming down to the inner basin. Trees in foreground at lower left are Engelmann spruce (*Picea engelmanni*).

San Francisco Mountain is about 10 miles north of Flagstaff, Coconino County. It is an eroded and truncated volcanic cone that earlier was about three thousand feet higher than it is now. Along the irregular crest line there are six peaks reaching above 11,000 feet elevation, and the mountain is known locally as the San Francisco Peaks. The three highest are Humphreys Peak (12,670 feet) on the northwest, Agassiz Peak (12,400 feet) on the southwest, and Fremont Peak (11,990 feet) on the south. One of the southernmost Pleistocene glaciers in the United States occupied the Inner Basin on the northeast side of the mountain, and was about two miles in length. The top of the mountain is about two miles in diameter and approximately a mile high, above its roughly 7,000-foot plateau base.

Elevation of the life-zone is from approximately 11,000 to 12,670 feet on Humphreys Peak and this is the alpine tundra community. Timberline is from approximately 11,000 to 11,400 feet, according to exposure. Precipitation is about 33 to 40 inches annually (it may well exceed 45 or 50 in any one year) and is distributed during both the cold and the warm seasons, but with relatively little occurring during the latter part of the growing season.

Fig. 51a. Alpine Tundra. Arctic-Alpine Life-zone. Easterly exposure below Humphreys Peak, *ca.* 11,600 ft., on the old Weatherford Trail, San Francisco Mountain, July 10, 1962; mid-July remnant of snow pack. Principal flowering herbs of foreground rock held are tundra daisies (*Erigeron simplex*) and gentians (*Gentiana* sp.). Small dwarfed trees appear as dark spots along ridge at upper left. Photo by R. R. Humphrey.

June to September is the usual growing season, with both the first frost and the first snowfall ordinarily occurring in early October. From November-December to March-April the mountain is snow-capped and patches of snow may remain into late August.

At timberline, the stunted and gnarled shrubby or prostrate Engelmann spruce and bristlecone pines grow singly or in scattered patches. Such patches may be found in alpine tundra beyond (above) the timberline tension zone (ecotone).[24] Two naturally shrubby species also occur here in the timberline transitional zone and just above it in wind-protected sites, *viz.*, gooseberry currant (*Ribes montigenum*) and dwarf juniper (*Juniperus communis montana*).

The complex of environmental determinants affecting the alpine timberline includes (1) strong ground wind forces and (2) winter low

[24] The "tree line" is theoretically that elevation of the last stunted tree, as distinguished from "forest line" which is the upper edge of the continuous forest.

temperature of air and soil, in relation to metabolism, photosynthesis, evapotranspiration, plant-available soil moisture, desiccation, wind shear, reproduction and cell death. Snow creep and avalanche, as well as tundra plant (sedge, genus *Carex*) competition, are additional known factors. Moreover, on the San Francisco Peaks above timberline the slopes are steep, rocky, and often covered with relatively loose volcanic debris. It is often difficult at best for trees to become established in such substratum conditions even without severe wind and frost action, and the treeline on the peaks is undoubtedly kept down at a somewhat lower elevation by the dual action of substratum and climate than it would be if it were a wholly climatic line (Brandegee, 1880; Merriam, 1899a; Shaw, 1909; Robinson, 1913; Adams, *et al.*, 1920; Forsaith, 1920; Pearson, 1920, 1931; Bates, 1924; Griggs, 1934, 1938, 1946; Shreve, 1942a; Daubenmire, 1943b; Raup, 1951).

Alpine tundra plants are uniformly small and low-growing, rarely more than a few inches above ground level, but many have large or showy flowers. Several are mat-forming cushion plants, and all are genetic dwarfs. Herbs, grasses, sedges, rushes, lichens, and mosses are characteristic; ferns and liverworts are also present. A remarkable number of the species in the flora are disjunctive from the high latitude tundra of Arctic North America and Eurasia; they range southward and eventually to the isolated high elevation alpine tundra in Arizona (San Francisco Mountain) and neighboring states, where these species reach their southernmost limits in North America. Twenty such species, roughly 40 per cent of the total of about 50, in the flora of the Arctic-Alpine Zone on San Francisco Mountain are such arctic-alpine disjuncts which also live in the arctic tundra. Fifteen of the twenty are circumpolar, *i.e.*, occurring in arctic Eurasia as well as in arctic North America. Two species appear to be endemic to San Francisco Mountain itself; a groundsel (*Senecio*) and a woodbetony (*Pedicularis*).

On the San Francisco Peaks only a few local areas above timberline are at all well covered with seed plants; this is largely a result of the steepness of the slope, looseness of soil, and presence of angular blocks and boulders. Some areas, such as unstable rock slides, are essentially without plants. The two primary tundra habitats and their plant associations are (1) the *alpine tundra rock field* (lichen association), and (2) the *alpine tundra meadow* (avens association). The two communities are often intermixed and many of the same species occur in both.

Alpine Tundra Rock Field. Crustose and foliose lichens occur on the rock surfaces, seed plants (herbs, grasses, sedges, rushes) are scattered between them, and mosses and ferns are found mostly in the rock crevices. The most abundant lichen is the crustose species *Rhizocarpon geographicum. Carex bella* is the common sedge, *Luzula spicata* (woodrush) the common rush, and the three grasses most commonly represented are alpine fescue (*Festuca ovina brachyphylla*), tundra bluegrass (*Poa rupi-*

cola), and spike trisetum (*Trisetum spicatum*). A single alpine fern, the bladder fern (*Crystopteris fragilis*) is commonly found in rock crevices. Most of the flowering plants above timberline are represented to some degree in this rock field habitat, including the most common of the species in the tundra meadow habitat (*Geum turbinatum* which is a mat-forming avens). The herbs of the rock field include several characteristic species which are arctic-alpine disjuncts, such as sandwort (*Arenaria sajanensis*), powderhorn (*Cerastium beeringianum*), moss campion (*Silene acaulis*), and Jacobs-ladder (*Polemonium confertum*).

Alpine Tundra Meadow. On San Francisco Mountain the alpine tundra meadows are not large nor particularly well-developed and they occupy but a small fraction of the Arctic-Alpine Zone. They are relatively "dry" and do not develop the "wet" meadow conditions as seen, for example, in the alpine tundra in the Colorado Rockies (Cox, 1933). The avens (*Geum turbinatum*) is the dominant plant, and other important mat-forming herbs include two arctic-alpine disjuncts, the circumpolar cinquefoil (*Potentilla sibbaldi*) and stemless catchfly (*Silene acaulis*). Two grasses essentially restricted to this association are mountain timothy *(Phleum alpinum)* and the bluegrass *Poa reflexa;* the other grasses listed above for rock fields are also present and are usually more abundant. In addition to the grasses, sedges (*Carex bella, C. albonigra, C. ebenea*) and rushes (*Luzula spicata, Juncus drummondi*) are, of course, typical of this association. Most of the herbs cited above for the alpine tundra rock fields are also present. In addition, there are a few herbs essentially restricted to the meadows and these include another arctic-alpine disjunct, the speedwell (*Veronica wormskjoldi*). Several species of mosses are present in the meadows, as is the one liverwort (*Lophozia porphyreleuca*) that is known from this zone on the mountain.

So it is, that in Arizona, isolated on high San Francisco Mountain above timberline, there live a number of circumpolar arctic-alpine plants that are widely distributed in arctic regions of the world such as Siberia, Alaska, and Greenland, and which extend southward on mountain tops of such impressive mountain ranges as the Himalayas, Alps, Sierra Nevada, and Rocky Mountains — and then reach the southern end of their distributions in North America on the high mountains of Arizona, California, and New Mexico.

With regard to the vertebrates, however, a single bird, the water pipit (*Anthus spinoletta alticola*) nests in the alpine tundra (meadow) on San Francisco Mountain,[25] and there are no mammals or other kinds of vertebrate animals represented (reproductively) there; man (*Homo sapiens*) is an occasional intruder. Three other birds which are well-known tundra species are represented in the faunas of New Mexico (white-

[25] The pipit also breeds in the alpine quasi-tundra on top of Baldy Peak (11,740 feet) in the White Mountains of Arizona, and in alpine tundra in New Mexico, but not in California.

tailed ptarmigan, brown-capped rosy finch) and California (gray-crowned rosy finch), in which regions there is more extensive development of the alpine tundra which remains today in the Southwest.

Boreal Life-zone[26]

The fir forest (Canadian Life-zone) and spruce-alpine fir forest (Hudsonian Life-zone) are frequently considered together as the Boreal Life-zone (boreal forest, Fig. 52), and, in its strictest sense, the Boreal Zone also includes the Arctic-Alpine Zone.

As noted above, the dominant plant life-forms of the Canadian and Hudsonian zones are similar. Also the two zones may merge gradually over wide areas and thus be made difficult to discern because of the partly azonal distribution of the dominant species; this is the problem of the continuum in the ecological distribution of plants and animals. The faunas of these two zones, and the Arctic-Alpine zone as well, are often quite similar in much of the Southwest.

In Arizona in particular there is a considerable intermingling of the vegetation dominants of the Canadian and Hudsonian zones as well as a considerable commonness of faunas, and there is but one species of vertebrate animal (pipit) peculiar to the small area of Arctic-Alpine. Thus it may be more convenient or meaningful for an investigator of some group of animals (vertebrate or invertebrate) to recognize a Boreal Zone; for example, in a biogeographic and evolutionary study of the Boreal Zone mammals of Colorado (Findley and Anderson, 1956).

The elevational range of the Boreal Zone in Arizona inclusive of the alpine tundra is from about 8,000 feet (7,500 feet on north-facing slopes) to 12,670 feet. In Arizona zonal precipitation is approximately 25 to 40 inches annually. Much of it is snowfall in these great western forest watersheds — the reservoirs that feed the permanent springs, streams, and rivers which traverse the arid and semiarid lands far below.

[26] Also *Boreal region.* The term *Austral region* was a collective term for Transition, Upper Sonoran or Austral, and Lower Sonoran or Austral; and Tropical region was a third term for "the tropical areas within the United States" and adjacent Latin America (Merriam, 1898:118).

Fig. 52. Boreal Forest. Boreal Life-zone. Spruce, fir, and aspen in the White Mountains. Looking northwest toward Baldy Peak (11,470 ft.) from a camp west of Big Lake, Apache County. The extensive light-colored stand of subclimax aspen in center is on a site of former disturbance (fire) within the forest. It is springtime and the heavy snowpack of winter is lingering on north-facing slopes. Photo by E. Tad Nichols.

SEQUENCE OF THE BIOTIC COMMUNITIES AND ZONES

From the foregoing discussion of Merriam's Life-zones, it is apparent that, in Arizona, the order of the zones from the lowest elevation (Lower Sonoran) to the highest elevation (Arctic-Alpine) is also the order of decrease in the total geographic area per zone. It may be noted also that Merriam selected names for the zones that are (with the single exception of Transition) quite appropriate in designating a general geographic region in western North America typified by the plant-animal communities represented in the particular zone; that is, the regions where the zones are best developed. When it is further noted that Arizona is closer to Sonora, Mexico than it is to the U.S.-Canadian boundary line, and even farther from the Arctic Circle, it is quite obvious why the extensive area of the two Sonoran Zones alone accounts for over 85 per cent of the total landscape of Arizona, and why the very small area of the Arctic-Alpine Zone is conspicuously less than even the relatively small area of the Canadian and Hudsonian Zones (boreal forest).

The Arctic-Alpine Life-zone occurs at sea level in Alaska and the Lower Sonoran Life-zone occurs at sea level in Mexico. Thus, when going upslope from desert to forest on a high southwestern mountain, one finds a progressive increase in "northern species" of plants and animals some of which extend to the Arctic Circle. Going downslope from forest to desert there is an increase in "southern species" some of which even extend to arid and semiarid lands in South America (see Floral Areas, below). In view of such obvious zonal relationships of life-forms with regard to latitude and "altitude" (elevation), one might expect quite correctly that some more or less valid generalizations have been formulated by biologists, climatologists, *et al.* of both the Old and New World (Humboldt, 1807, 1820; Merriam, 1890; Robbins, 1917; Hopkins, 1918, 1938; Vischer, 1924; Haviland, 1926; Cain, 1944; Good, 1947; Dansereau, 1957; and many others). Merriam discussed the "law of latitudinal equivalent in altitude," noting its early formulation by Alexandre de Humboldt, and he first presented some data and estimates for the Colorado Plateau with special reference to the San Francisco Mountain region in Arizona.

The general proposition, or law of the latitudinal equivalent in elevation, is stated today essentially as it was well over a century ago. That is, that the temperature decrease (lapse) from the equatorial zone to the poles occurs at the average rate (lapse rate) of *ca.* 1° F. for each 1° of latitude; and decreases from a lower to a higher elevation in the same general area at the rate of *ca.* 1° F. for each 330 feet of elevation.

This generalization for the world is made with full cognizance of variations due to regional topography, local influences, and seasonal differences, all of which, of course, affect more than just the temperature gradient. In Arizona the overall (mean annual) thermal lapse rate is close to 4° F. (2.2° C.) per 1000 feet (305 meters) elevation, *i.e.*, approximately 1° F. per 250 ft. (Merriam, 1890; Shreve, 1915; Pearson, 1920, 1930, 1931; Green, 1961). Shreve (1915) found the mean temperature gradient to be 4.1°F. per 1000 ft. on the Santa Catalina Mountains in southern Arizona during 1908-1914. Pearson's data (1920) for 1917 and 1918 show a similar lapse rate of 3.7°F. on a gradient in northern Arizona from desert-grassland at Kingman (3,300 ft.) to timberline (11,500 ft.) on San Francisco Mountain; see Table 3, page 16.

Table 4. Mean Annual Precipitation in Biotic Communities in Arizona and New Mexico. Data from Pearson (1931).

Biotic Community	Inches
Spruce-alpine fir Forest	34
Douglas fir Forest	26
Ponderosa pine Forest	21
Juniper-pinyon Woodland	17
Grassland	11
Desert	10

Temperature inversions result from local disturbing influences that often reverse otherwise linear temperature gradients in mountainous country. It is well known that cold air tends to drain off of high places such as mountain slopes, and to settle in low places such as valleys. Nocturnal downslope cold air drainage causes temperature inversions such as those seen in the reversal in the *minimum* temperature data in Table 3 for the ponderosa pine forest below (colder) and fir forest above (warmer) on San Francisco Mountain; the ponderosa pine zone actually has a shorter frost-free season. Marked temperature inversions are common occurrences in the Rocky Mountain region. Thus milder "thermal belts," of a few hundred vertical feet, often lie between colder air both below and above. It may be surprising but as a general rule, in the Rocky Mountain region, inversions take place during the night at most localities during all seasons of the year.

The greatest distance south to north across Arizona is 392 miles; this is equivalent to 5.68 degrees of latitude. Thus in terms of the law of latitudinal equivalent in elevation and using the relationship for Arizona of 1 degree of latitude per 250 feet elevation, this S-N distance alone may be comparable in its effect on prevailing air temperature to an elevation change of about 1,400 feet. Also, if one assumes the phenological rate of approximately four days equivalence to each 1 degree of latitude in temperate North America (Hopkins, 1918), the spring events in the seasonal biological advance in extreme northern Arizona lag by about 3 weeks those occurring at the same elevation in extreme southern Arizona.

Soil temperature decreases as does the air temperature with elevation. The last killing spring frost is 10-15 days later for each upward 1000 feet, and the frostless season decreases—on the order of 15-30 days per 1000 feet elevation—from the southern desert (40 or more weeks) to forest and tundra (about 15 weeks) according to latitude and locality (Shreve, 1915, 1924; Cannon, 1916; Pearson, 1920; Smith, 1956; Sellers, 1960a).

Elevation ("altitude") is of far greater importance than change in latitude in shaping the character of climate, vegetation, and the resulting biotic communities in Arizona. Coupled with this great variation in height is the remarkably varied topographic variation in space. On elevation

Table 5. Average daily evaporation (E), soil moisture (SM), and the ratio of evaporation to soil moisture (E/SM) for south-facing exposures at six elevations in the Santa Catalina Mountains in southern Arizona during the arid fore-summer of 1911. Data from Shreve (1915).

Habitat	Elevation (Ft.)	Evaporation (E)	Soil Moisture (SM)	E/SM
Forest	8,000	29.3	7.4	3.9
	7,000	62.8	2.6	24.1
Woodland	6,000	59.4	1.8	33.0
	5,000	61.7	3.1	19.9
Desert-Grassland	4,000	80.4	2.0	40.2
Desert	3,000	101.1	2.0	50.5

gradients environmental factors operate together and affect the actual elevational limits of plant and animal species, and hence the sequence of zones and the biotic communities which comprise them. Such major physical factors of the environment in addition to light and soil, are temperature, precipitation, evaporation, humidity, cloudiness, and wind (Tables 3, 4, and 5).

The direction of slope exposure is a powerful indirect factor which causes constant departures of biotic communities from the normal gradient. Just as the lower elevational limit for a species is at a lower elevation on a north-facing slope, so it is that it occurs at still lower elevations on such slopes in the northern part of the state (or its range) than in the southern part. In addition to the steepness of slope and the important effect of slope exposure, the following are some of the often important local influences which interfere with the normal elevational sequence of life-zones (Merriam, 1891; Shreve, 1915; Standley, 1915; Hall and Grinnell, 1919; Jepson, 1925): (1) extent of mountain mass as well as height and base-level (the *Merriam Effect;* Merriam, 1890; Grinnell and Swarth, 1913; Shreve, 1922; Marshall, 1957; Lowe, 1961; Martin, *et al.*, 1961), (2) air currents, (3) cold water streams and drainageways, (4) moist soil evaporation, (5) proximity of large bodies of water, (6) snow banks and glaciers, (7) rock and rocky surfaces, and (8) miscellaneous local influences such as fires and other denuding agents including man, his implements and his domestic animals. All of these influences can in some measure affect a zonal habitat interdigitation that can be seen somewhere on any high mountain in the West. Such downward interfingering of forest and woodland species, however, takes place within narrow canyons and valleys rather than in wide valleys; within wide valleys species of lower zones extend farther upward as they also do on the ridges.

Some of the local influences noted above can produce a *zonal inversion* as can be seen on and directly below the south rim of the Grand Canyon where stands of Douglas fir occur below the pinyons, junipers and ponderosa pines which are on the rim. And even the complete absence of one or more zones may occur in some mountains.

As already noted, the total precipitation (rain and snow) increases while the temperature decreases with increase in elevation. Because of this, the maximum of one factor coincides with the minimum of the other. Consequently, in Arizona the animals and plants with *high moisture* requirements (*e.g.*, spruce trees) must be able to, and do, make a successful living in low temperature environments such as forest or tundra, and those with *high temperature* requirements (*e.g.*, reptiles) must be able to, and do, make a successful living in environments with relatively little moisture, such as desert and grassland. The *upper* elevational range of plant species and animal species on mountains in Arizona is determined by their ability to function at low temperature. The *lower* range limit of woodland and forest species is determined by their ability to resist drought — and for the plants this means the plant-available soil moisture.

The extremes rather than the means of the daily weather elements (and climate) are critical. This is an extension of Liebig's law of the minimum (see Odum, 1959). It is indeed interesting that the upper elevational limits are determined by low temperature and that (1) for forest species, this deficiency of heat manifests itself in *low maximum* temperatures rather than in low minima, while (2) for desert species, *e.g.*, Sonoran Desert, this deficiency of heat manifests itself in the *low minima* of winter (usually in January, Fig. 53). The rapid decrease of physiologic temperature efficiency with increase in elevation (Pearson, 1931) is associated with the remarkably greater frost-hardy adaptedness of the high-elevation genotypes. In Arizona, death from winter-killing by low temperature has relatively little effect on forest (and grassland) species derived from the Arcto-Tertiary Geoflora. However, winter-killing has a very pronounced effect on the tropically related desert species derived from the Madro-Tertiary Geoflora (Shreve, 1911, 1915, 1922; Thornber, 1911; Turnage and Hinckley, 1938; Detling, 1948; Lowe, 1959a). Duration of the temperature in hours or days is often as important a temperature factor as the degree of temperature itself, especially in the case of the extremely low winter minima (Shreve, 1911, 1915).

The average (mean or median) rate of change of precipitation (4-5 inches per 1000 ft.) on typical mountain gradients, or anywhere else in Arizona, is of much less importance than knowledge of the actual variation (extremes) of the rainfall conditions from year to year at a given locality and the effects these have on the plant-available soil moisture. Moisture factors in addition to the precipitation are humidity, cloudiness, wind, evaporation, transpiration (plant), and the all-important resulting soil moisture. Humidity, cloudiness, wind, and temperature are the joint determinants of the rate of evaporation, and evaporation (or evapotranspiration), like temperature, decreases with elevation. The ratio of the evaporation to soil moisture expresses the critical conditions which determine the lower limits of plants on moisture gradients in the Southwest, and these are tremendously different ratios for forest and desert (Table 5). "The ratio of evaporation to soil moisture comprises a measurement of all the external factors which affect the water relations of plants, except the influence of radiant energy on transpiration and the possible effects of soil temperature on this function. It is accordingly unnecessary to give further consideration to rainfall, which is not in itself a factor for vegetation in such a region as Arizona." (Shreve, 1915, page 93.)

Soils in Arizona from desert to forest appear to exert their very important influence on the nature of the biotic community through their physical rather than their chemical characteristics; and by far the most important effect that variation in soils has on all of the biotic communities of Arizona is the direct effect such variation has on the increase or decrease of the plant-available soil moisture. Other things being equal, the more porous soils are best suited for tree growth. In the desert (1) the physical composition of the soil (rock, sand, silt, clay) and (2) its depth are the primary limiting characteristics for both plants and animals. In the forest only the organic content is a major addition to these two characteristics (McGee and Johnson, 1896; Shreve, 1915, 1924, 1951; Pearson, 1920, 1931; Nikiforoff, 1936; Martin and Fletcher, 1943; Marks, 1950; Yang and Lowe, 1956). Desert species reach highest elevations on limestone, next highest on volcanics, and lowest on gneiss (Shreve, 1919).

From the discussions above it is evident that we are particularly indebted to Shreve and to Pearson, as well as to Merriam and others, who were among the early scientific pioneers of environmental gradients on

Fig. 53. Sahuaros (*Cereus giganteus*) killed during extreme freeze of January 12-13, 1962. Paloverde-sahuaro community, 3,200 ft., on rocky slope south side Santa Catalina Mountains. Over 35 per cent of stand in photo was killed; death of up to 70 per cent occurred in some stands. Survivors possess genotypes bestowing greater "cold-resistance." Photo by Richard D. Krizman.

high, rugged mountains in Arizona. They well-documented vegetation gradients on slopes as controlled by environmental moisture and temperature gradients — often steep slopes and gradients from the desert directly up into the forest, with greater environmental change than that between the states of Florida and Maine. In so doing they were documenting the moisture-temperature gradients which control the biotic communities over much of the Southwest. It is significant that Shreve (1915, and elsewhere) working in southern Arizona, and Pearson (1920, and elsewhere) working in northern Arizona arrived at the same conclusions with regard to the correlation of vegetation with the upslope increase in moisture and decrease in temperature and evaporation. That is, as noted above, that (1) the lower elevational limit of a given species of plant on the moisture-temperature gradient is controlled by deficient moisture, and (2) that the upper elevational limit is controlled by deficient heat. It is beyond reasonable doubt that these findings are correct (Daubenmire, 1934a, 1959; Linsey, 1951).

Just as the environmental gradients on mountain slopes are essentially continuous (although variable) linear rates of change, so it is that organisms (plants and animals) are distributed in a more or less continuous type of pattern, *i.e.*, a *continuum* which is more easily seen in the climax pattern of the plants (vegetation) than in that of the animals. Each species of plant and animal, responding on its own to the limits set by its genotype, will "come in" at some point on the gradient and gradually reach its metropolis before finally disappearing at some point still farther along the gradient (Shreve, 1915; Whittaker, 1956, 1960). But a single species of plant sets the limits to the Pacific redwood forest (the coast redwood) just as a single species outlines the ponderosa pine forest (the ponderosa pine), etc. (Mason, 1947).

Such predominant plants and others, and some animals, have been called indicator species, life-zone indicators, etc. (Merriam, 1898; Bailey, 1913; Hall and Grinnell, 1919; Clements, 1920). While "indicators" as such are little discussed any more, they are nevertheless inherent in the thinking and mapping of natural ecologic units such as biome, life-zone, province, type, etc. Witness the boundary between the Great Basin desertscrub and the Southwestern desertscrub *(i.e.,* the northern boundary of the Mohave Desert) drawn on the basis of the northern continental limit (in Nevada and Utah) of but a single species of plant — the creosotebush, *Larrea divaricata.* And when it is stated that "A subalpine zone begins at about 9,500 feet" in Utah, it is a statement of fact concerning the distribution of but a single species of forest plant, Engelmann spruce (*Picea engelmanni*). Moreover, the entire coniferous forest itself in the Inland Southwest "comes in" on the moisture gradient with the first individuals or stand of the one species of pine (ponderosa) at the lower limit of the ponderosa pine forest; this is ordinarily between 6,000 and 7,000 feet elevation.

Figure 54 depicts the continuum of pines and oaks on southerly exposures on the south-facing side of the Santa Catalina Mountains, "facing" the Sonoran Desert. Note that only one plant (Emory oak) approximates the limits indicated for the evergreen woodland and that no two species of oak have the same span of ecologic tolerance. Note that only one plant (ponderosa pine) approximates the limits indicated for the pine forest and that no two pines nor trees on the mountain slope coincide in all of their distributional limits. All of these dominant and subdominant species, shown occurring over a vertical distance of 5,000 feet, are in a woodland and forest continuum which begins below as an "open encinal" or oak-grass landscape between 4,000 and 4,500 feet, and increases upslope in density and in cover on the moisture-temperature gradient to an oak-juniper-grass landscape between 4,500 and 5,500 feet, and with increased density, coverage, and floral components to an oak-juniper-pine landscape (oak-pine woodland) that merges with the ponderosa forest above.

The vertical spans of ecological tolerance shown in Figure 54 represent the ecologic summations of the integrated physiological, behavioral, and morphological tolerances of each species. The overlap of these genetically controlled tolerances is the necessity which naturally precedes and permits all of the important biotic relationships which can then be, and are, subsequently established among plants and animals. In the West, and particularly in the desert, the genetic or "individualistic" relation of the plant to environment is far more striking, and more important, than the "organismic" relation of plant to plant. Here the biotic communities that comprise the climax patterns owe their ultimate existence first and decidedly foremost to the coincidental overlap of the genetically controlled spans of tolerance of the comprised species in relation to the environmental gradients and mosaics.

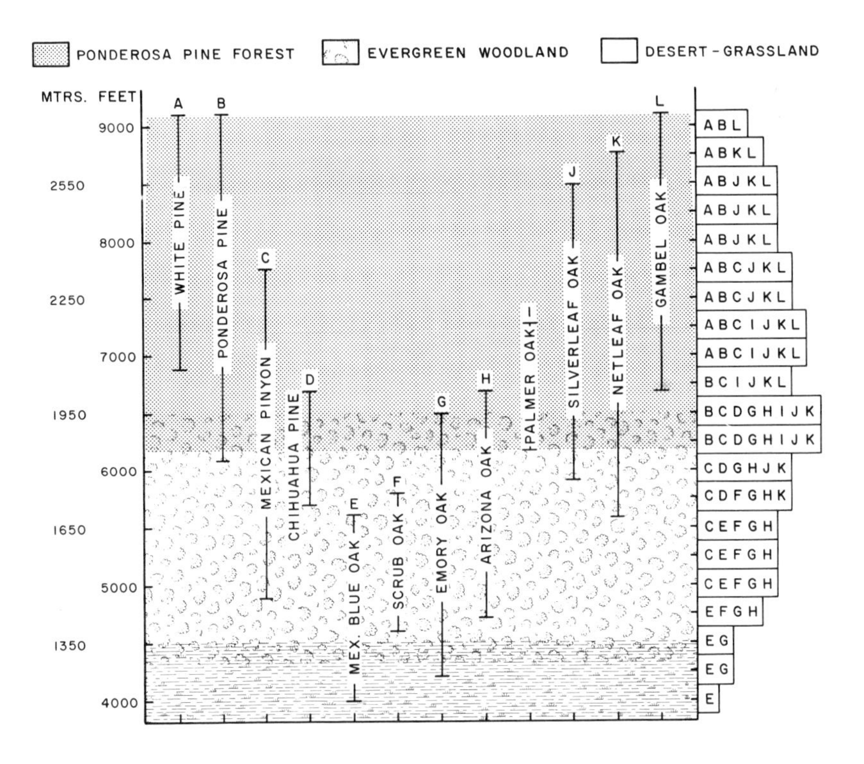

Fig. 54. Amplitudes (spans) of ecologic tolerance for the oaks and pines in the woodland and forest continuum and mosaic on south-facing slopes on the south side of the Santa Catalina Mountains. The overlapping vertical lines indicate the continuum of species frequency distributions; the zipatone background indicates major components of the mosaic. See text.

BIOMES

Some of the American animal-oriented ecologists would prefer to add the term *biome* as an additional label to the major ecologic formations. Thus, for example, instead of speaking of a species that is a forest bird or a grassland bird, one might say a forest biome bird or a grassland biome bird, etc. The term "biome" is redundant. It has most often been used with the purpose of emphasizing the centuries-old and well-known

Table 6. The Ecologic Formations in Arizona and their Equivalent Biome (Shelford) and Life-zone (Merriam) Designations

Ecologic Formations* Lowe, 1961	Biomes* Shelford, 1945	Life-zones Merriam *et al.*, 1910
Desert	. . . . .	. . . . .
Southwestern Desertscrub (Mohave, Sonoran, and Chihuahuan Deserts)	Creosotebush-Kit Fox (Hot Desert)	Lower Sonoran
Great Basin Desertscrub (Great Basin Desert)	Shadscale-Kangaroo Rat (Cool Desert)	Upper Sonoran
Grassland	Gramma Grass-Antelope (Grassland)	Upper Sonoran
Desert-Grassland	. . . . .	" "
Plains Grassland	. . . . .	" "
Chaparral	*Adenostoma*-Brush Rabbit† (Chaparral)	" "
Interior Chaparral	. . . . .	" "
Woodland	. . . . .	" "
Oak Woodland	. . . . .	" "
Pine-oak Woodland	. . . . .	" "
Pinyon-juniper Woodland	Juniper-Rock Squirrel (Pinyon-Juniper Woodland)	" "
Coniferous Forest	Spruce-Moose† (Coniferous Forest)	Boreal (part)
Pine Forest	. . . . .	Transition
Fir Forest	. . . . .	Canadian
Spruce-fir Forest	. . . . .	Hudsonian
Tundra	*Cladonia*-Caribou†	Arctic-Alpine
Alpine Tundra	. . . . .	"

*Riparian woodland, encinal, and mountain grassland are not listed in the left-hand column of this table; they also lack Shelfordian biome equivalents.

†The names *Adenostoma, Cladonia,* Moose, and Caribou refer to native plants and animals which, of course, do not occur in Arizona, other southwestern states, or in adjacent Mexico.

fact that animals and plants live and function together in natural communities; "biome" is synonymous with ecologic (=biotic) formation and climax (see Table 6) in practice (if not always in theory; see Tansley, 1935). And, as mapped, "biome" is synonymous with climax "plant formation," regardless of the animals present. Thus the major biotic community of associated plants and animals that is the plant-animal formation which we recognize and call *coniferous forest*, has been called the spruce-moose biome, coniferous forest biome, and other "biomes" (see, for example, Shelford, 1932, 1945; Clements and Shelford, 1939; Pitelka, 1941; Allee, *et al.*, 1949; Kendeigh, 1961).

Kendeigh (1948, 1954, 1961) recently has made an interesting attempt to further justify the addition of a still more varied biome terminology to our already somewhat large ecological vocabulary. It may be doubtful, however, that the recognition and study of plant-animal organization, succession and evolution requires the additional ecological jargon (biocies, biociation, biome, biome type, *et al.*). The prior ecological term *formation* (Grisebach, 1838; Weaver and Clements, 1938; and Clements and Shelford, 1939) and its straightforward application in the obvious sense of an ecologic formation (Table 1), *i.e.*, in the only true ecologic sense of a biotic (plant-animal) formation, seems unnecessarily extended to biome, etc.

The paper by Rasmussen (1941) on biotic communities of the Kaibab Plateau in northern Arizona should be consulted. This is a good study carried out under the direction of V. E. Shelford and is found with some surprise to be free of biome jargon. Rasmussen studied both the plants and animals of major biotic communities (forest, woodland, desert) and rejected the frankly annoying implications (of Shelford, Kendeigh, *et al.*) that biome jargon is necessary or even useful for the study and understanding of biotic communities.

BIOTIC PROVINCES

Whereas "life-zones" and "biomes" are fully biogeographic systems, *biotic provinces* are usually, though not always (see Munz and Keck, 1949), more faunistic in their conception. They are continuous geographic areas arbitrarily delimited by various means. Dice (1943: 3) mapped biotic provinces for North America and defined them as follows: "Each biotic province covers a considerable and continuous geographic area and is characterized by the occurrence of one or more important ecologic associations that differ, at least in proportional area covered, from the associations of adjacent provinces."

This definition is, of course, at best a loose one and points to the complexity of criteria often employed, and the obvious subjectivity resulting when they are applied. Vestal (1914) suggested that a biotic province involves similarity in the geographic ranges of animals of ecological similarity, as well as close correspondence between animals and vegetation. Munz and Keck (1949: 88) regard California as a region which "may be divided naturally into a few [5] major biotic provinces, as determined by broad differences in climate." Miller (1951:581) states for biotic provinces, that "Their only essential features seem to be some distinctness of their faunas. Barriers, whether zonal, biotic, or physiographic, are the critical agents that set off an area and its fauna and keep it partly different from that of an adjoining area."

Whatever the criteria employed, and however limited the animal groups emphasized (*e.g.*, small mammals, large mammals, all mammals, or all vertebrates), biotic provinces are concerned with, and usually are intended to show, subcontinental or smaller regions (often quite small) of faunal differentiation (subspecies, species, genera).

By indicating geographic areas of evolutionary differentiation, biotic provinces are not necessarily different from life-zones, biomes, and faunal areas. Biotic provinces are different from life-zones and biomes, however, by being continuous geographic areas, and also by being primarily faunal concepts and representations rather than those of biotic communities. The system may be useful in ecological and evolutionary studies inasmuch as certain regional relationships and differences may be revealed (*e.g.*, Merriam, 1890; Ruthven, 1908; Gates, 1911; Vestal, 1914; Van Dyke, 1919; Dice, 1922, 1939, 1943; Blair, 1940, 1950, 1952; Smith, 1940; Moore, 1945; Goldman and Moore, 1946; Jameson and Flury, 1949; York, 1949; Miller, 1951; Peters, 1955; Denyes, 1956; Munz and Keck, 1959).

Inasmuch as the biotic province was originally a primarily faunistic conception in North America it is not surprising that the more recent as well as early literature on the subject in this country has been largely associated with the investigation of the ecology and distribution of animals rather than plants. Biotic provinces and their uses (and/or misuses) have been discussed recently by Pitleka (1943), Parker (1944), Blair and Hubbell (1938), Allee, *et al.* (1949), Miller (1951), Kendeigh (1954, 1961), Clarke (1954), Woodbury (1954), Peters (1955), Cain and Oliviera Castro (1959), Munz and Keck (1959), and others. The essentials of the concept and its use are well presented most recently by Cain and Oliviera Castro (1959) in their manual of vegetation analysis. Patterns and criteria of distribution in general are recently discussed by Dansereau (1957) and Darlington (1957); see also Hesse, Allee, and Schmidt (1937), Cain (1954), and Beaufort (1951).

C. Hart Merriam (1890) was the first to recognize and name biotic provinces in Arizona. He outlined a "Sonoran province," a "Great Basin province," and a "Boreal province." Dice and Blossom (1937) recognized two biotic provinces in southern Arizona, the "Sonoran province" and "Chihuahuan province." Dice (1939) discussed this further, and named and arranged provinces for Arizona essentially as later mapped (Dice, 1943). These are as follows (Fig. 55; compare with Fig. 1 and Fig. 56):

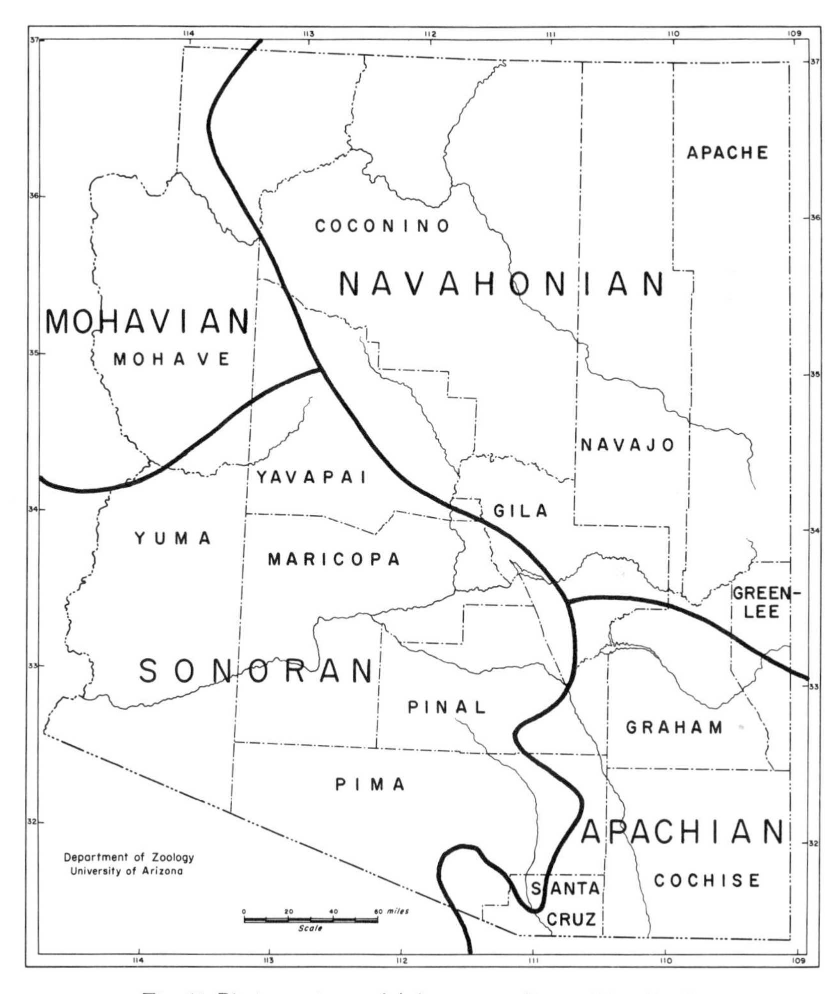

Fig. 55. Biotic provinces of Arizona according to Dice (1943).

1. The *Sonoran biotic province* is desert, and, in Arizona, it is intended to be essentially the area of the state that is within the Sonoran Desert.

2. The *Mohavian biotic province* is essentially the Mohave Desert of other authors (*sensu* Shreve, 1942).

3. The *Apachian biotic province,* according to Dice, is intended to represent the "grassy high plains, and the mountains included in them, of southeastern Arizona, southwestern New Mexico, northeastern Sonora, and northwestern Chihuahua." This is the Yaquian biotic province of others.

4. The *Navahonian biotic province* (Dice, 1943:39), "... is characterized by pinyon-juniper woodland." The "lowest life belt" of the Navahonian is characterized by "arid grassland," and the highest, the "alpine life belt," is made up of treeless areas above timberline. Dice does not mention the Great Basin Desert in his discussion of this province; the southwestern corner of his *Artemisian biotic province* (in the Great Basin) lies in southern Utah.

Dice (1922), Clark (1937), and Davis (1939) have used the term "biotic area," and Allen (1892) used the term "life area," to distinguish smaller plant-animal "units" within a still greater area, say, of the size of the state of Arizona, or of Idaho (Davis, 1939).

FAUNAL AREAS

The essentially, if not purely, faunistic concept and term *faunal area* antedates those of life-zone, biome, and biotic province in North America. Faunal areas are not biotic provinces (in the sense of Dice and Vestal), although they may be similarly conceived and are often equally subjective and difficult to define. Faunal areas may or may not have similar plant associations, but they have dissimilar faunas. Both before and since Allen's use of the term faunal area (1871, 1892, 1893) for the smallest units in his classification and analysis of North American vertebrates (for an area with certain species more or less restricted to it), several investigators have reported on "faunal areas," "faunal districts," etc. of the Southwest (Le Conte, 1859; Cope, 1866, 1875; Packard, 1878; Merriam, 1892; Mearns, 1896, 1907; Townsend, 1897; Brown, 1903; Stevens, 1905; Ruthven, 1907; Grinnell, 1914, 1915; Howell, 1923; Bancroft, 1926; Grinnell, 1928; Law, 1929; Swarth, 1929; Gloyd, 1932, 1937; van Rossem, 1936a; Burt, 1938; Phillips, 1939; Johnson, *et al.*, 1948; Webb, 1950; Miller, 1951; and others; see Law, 1929, for an early critique).

Gloyd (1937) investigated herpetofaunal areas in southern Arizona, mapped them and showed their relationships to the concepts of Mearns (1907) and Swarth (1929). See Figure 56 from Gloyd (1937, Fig. 11) and compare it with Figure 1 (Sellers, 1960b) and Figure 55 (Dice, 1943). Van Rossem (1936a) reached the same general conclusions (based on bird distribution) as those of Gloyd (based on reptile distribution) and further advocated the Baboquivari Mountains rather than the Santa Rita Mountains (Swarth, 1929) as the western boundary of the "Eastern Plains Area."

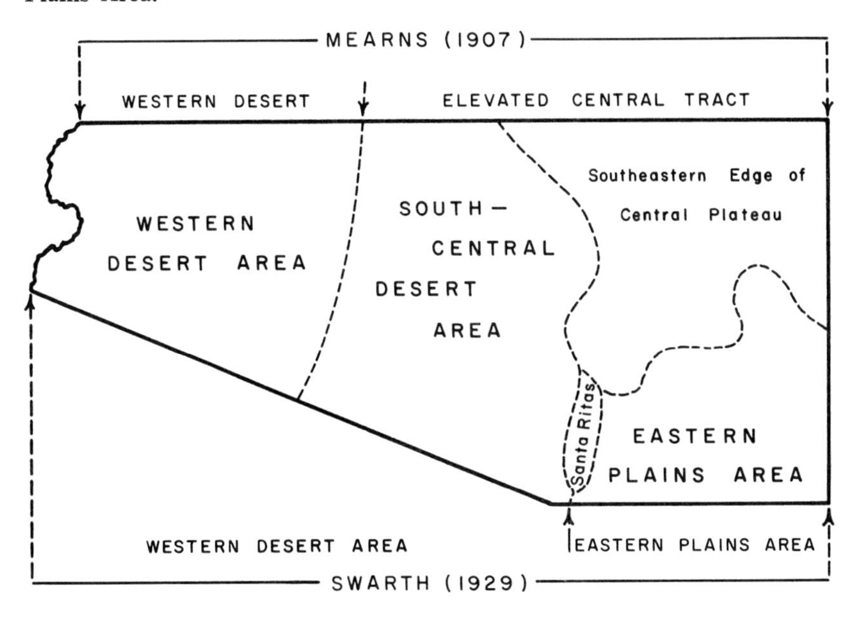

Fig. 56. Faunal areas of southern Arizona, based on the distribution of reptiles, and comparison with the concepts of Mearns (1907) and Swarth (1929); from Gloyd (1937).

Phillips (1939) studied avifaunal areas, reviewed much of the pertinent literature, and recognized four such areas in the state, as follows: "Yuman faunal area," "Pima faunal area," "Mogollon faunal area," and "Navajo faunal area." Phillips and Monson (1936) have also designated "faunal regions" of Arizona based primarily on the geographic distribution of breeding birds.

Faunal areas (faunal districts, faunal territories) are geographic areas or regions where the distributional ranges of several animal species are similar, and where they reach their maximum occurrence. Within the region there are usually a number of the species which are characteristic of other "foreign" faunal areas. Thus a fauna may consist of one or more abstract *faunal elements* such as a Rocky Mountain element, Sierra Madrean, Chihuahuan, Sonoran, Great Basin, Great Plains, Californian, and other elements, which is ordinarily the case for Arizona and other Southwestern "faunas" (see Miller, 1951).

FLORISTIC ELEMENTS

Floras, floristic areas, floristic territories, floristic elements *et al.*, are the plant counterparts in biogeography (floristic geography; see Cain and Castro, 1959). The concept of *floristic element* (=geographic element) as well as of vegetation is important to an understanding of the history and evolution of our present southwestern landscapes (see Axelrod, 1950, 1958).

Arizona and New Mexico, with adjacent parts of Texas and California, sit geographically at the terminus of a host of cool-northern American species which extend this far southward on the continent, and a host of warm-subtropical American species extending this far northward. They meet here and variously co-mingle, in this region called the Southwest, in a complex pattern which is part of a continuum on a subcontinental as well as historical scale. The warm-subtropical species comprise derivative communities of the Madro-Tertiary Geoflora (Southwestern desertscrub, desert-grassland, plains grassland, chaparral, evergreen woodland); the cool-northern species comprise derivative communities of the Arcto-Tertiary Geoflora (alpine tundra, coniferous forest, mountain grassland, Great Basin desertscrub).

This is clearly revealed on a "desert mountain" such as the Santa Catalinas in Arizona. The flora in particular reveals three great segregations of the biota during the Pliocene-Pleistocene into (1) an older *northern* (forest and tundra) element now restricted during the current post-Pleistocene to the higher elevations on the higher mountains, (2) a younger *southern* mixed element at mid-elevations (woodland and chaparral), and (3) our newest, youngest environment existing today — desert — occurring at the lowest surrounding elevations, with its vegetation, flora, and fauna derived through evolution from *southern subtropical* and tropical stocks chiefly during, and since, the Pliocene.

Shreve (1915: 40) discussed the floristic elements in the desert, woodland, and forest vegetation of the Santa Catalina Mountains (to 9,150 feet) in southern Arizona and pointed to several related problems involved in ecology and evolution:

> "To summarize for the mountain as a whole, it may be said that the floristic relationships of the Desert and Encinal [Woodland] regions are almost wholly with the Mexican deserts and foothills to the south, while those of the Forest region are divided between the Mexican Cordillera and the Rocky Mountains. The Mexican group is the more conspicuous in the make-up of the vegetation, while the Rocky Mountain contingent is apparently preponderant in number of species."

The following breakdown of the vegetation and floristic elements in the Santa Catalina Mountains is essentially from Shreve (1915).

Desertscrub

The specific and generic relationships of the desert flora (Sonoran Desert) of the Santa Catalinas are, in over 90 per cent of the cases, with Latin America. That is, the relationships are with the Mexican deserts (Sonoran, Chihuahuan, Tehuacan) and semiarid and subtropical lands, the South American deserts and subtropical lands of Argentina and Chile, and to a lesser extent with the Caribbean and Tropical American regions. There is, on the other hand, little relationship with the northerly Mohave Desert and virtually none with the still more northerly Great Basin region, except for some spring annuals. Just the opposite is true for mountain vegetation in *northern* Arizona. Moreover there is relatively little true endemicity (Mogollon element) in the flora of the Santa Catalinas, or, in fact, anywhere in the state.

Grassland

The desert-grassland is of fairly extensive development on all sides of the Santa Catalina Mountains except on the south. On the south side of the mountain the desert-grassland is compressed into a narrow vertical zone of roughly 500 feet elevation on steep gradients at the upper edge of the desert between about 3800-4300 feet elevation where it is confluent with a brief but gradual transition (ecotone) from desert into woodland. The floristic elements are essentially four: Great Plains, Chihuahuan, Sonoran, and Sierra Madrean.

Woodland

The floristic elements in the foothill and mid-elevation evergreen woodland vegetation of the Santa Catalinas are predominantly Madrean; that is, Chihuahuan and Sonoran in the sense of being mid-elevation Sierra Madrean species on both sides of the Cordillera. There is also a Californian element as well as Great Plains, Rocky Mountain, and Eastern North American elements.

Forest

The conspicuous plants in the coniferous forest of the Santa Catalina Mountains (pine forest and fir forest) are primarily Rocky Mountain and Sierra Madrean species. Others represent a Boreal element as well as Californian, Pacific Northwestern, Eastern North American, Mogollon (essentially Arizona-New Mexico), and Cosmopolitan elements.

Shreve (1915:40-41) concluded that:

> "It will be impossible to summarize the floristic relationships of the Santa Catalinas in a thorough manner until very much more is known of their own flora and also of the floras of the many adjacent mountain ranges and desert valleys, both in the United States and Mexico. For the explanation of these relationships a closer acquaintance is needed with the actual mechanisms of transport which are effective in the dispersal of the seeds of desert and mountain plants. *A fuller knowledge is also required of the fluctuations of climate within recent geological time, and of the consequent downward and upward movements of the Encinal and Forest belts of all southwestern mountains* [italics mine]. Such movements would alternately establish and break the connections between the vegetation of the various mountain ranges and elevated plains, thereby permitting the dispersal and subsequent isolation of species which might find no means of movement across the desert valleys under existing conditions."

Shreve wrote this at the former Carnegie Desert Laboratory, on Tumamoc Hill at Tucson. Today Shreve's many far-sighted leads in environmental biology are being enthusiastically followed by several investigators at the University of Arizona where, for example, in the same buildings on Tumamoc Hill, the Geochronology Laboratory of the University is investigating physical, biological, and cultural aspects of arid lands (see "The Last 10,000 years" by Martin, Schoenwetter, and Arms, 1961).

AQUATIC HABITATS

The aquatic fishes and semi-aquatic frogs, turtles, beavers, and other "amphibious" species in Arizona live in either warm-water or cold-water rivers, streams, lakes, or ponds. The major localities in Arizona's warm-water and cold-water fisheries are shown on maps provided in Mulch and Gambel's useful pamphlet (1954) on game fishes in Arizona.

Figure 57 is a drainage map of Arizona (Miller, 1951) which includes those waters supporting permanent populations of native fishes. While there was additional disturbance of some of these waters during the past decade, they remain today largely as shown. The major drainage basins in Arizona are outlined on this map, which also accompanies the check list of fishes (Miller and Lowe, part 2 of this work, 1964).

Figures 58 and 59 show parts of the Colorado River, Arizona's largest, as it courses through the deep gorge that is Grand Canyon. The aquatic fauna of this sizeable Arizona section of the Colorado, which drops 2,000 feet in elevation from *ca.* 3,200 at Lee's Ferry in the Great Basin Desert near the Utah line to *ca.* 1,200 at Hoover Dam (Boulder Dam) in the Mohave Desert in Nevada, is an immensely interesting fauna that has been little studied and is consequently little known.

Figures 65 and 66 are photographs of Aravaipa Creek, a desert stream in the Galiuro Mountains of southern Arizona that is partially canopied by a well-developed riparian woodland climax dominated by cottonwood, willow, and ash (cottonwood-willow gallery association). The desert

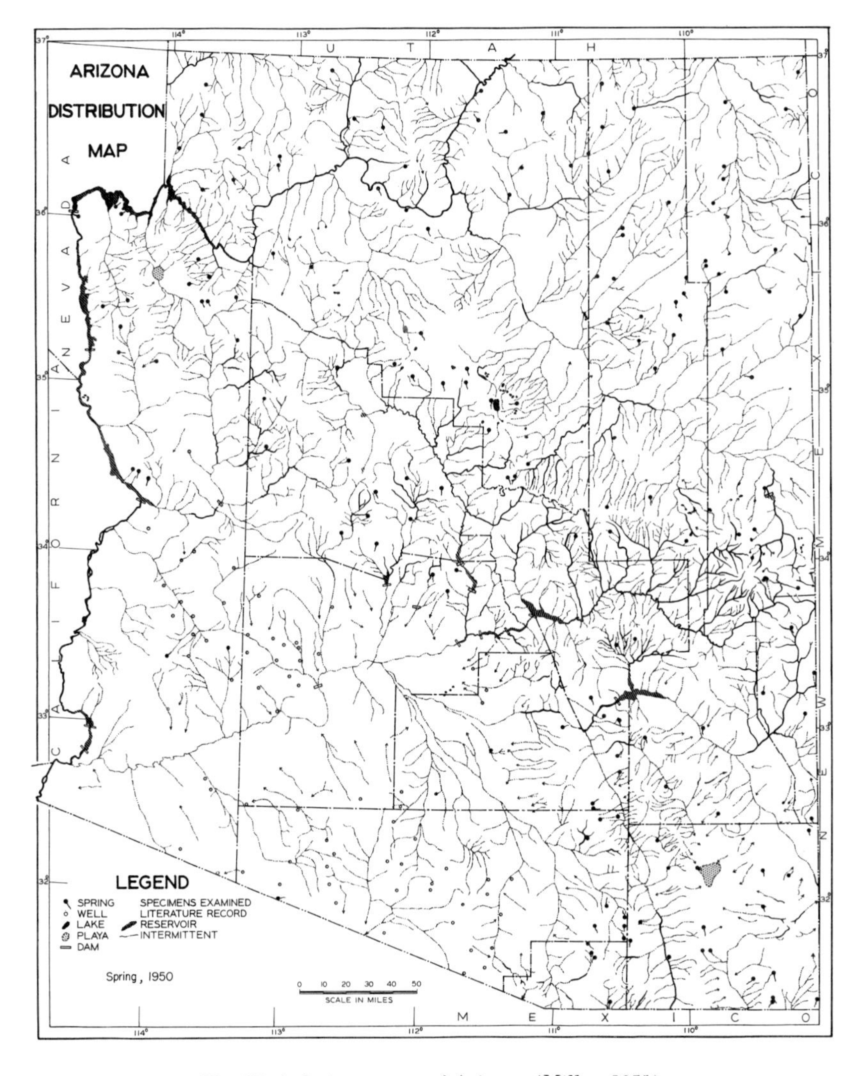

Fig. 57. A drainage map of Arizona (Miller, 1951).

climax here, at 2,600 feet elevation, is paloverde-sahuaro. This fast-flowing, spring-fed stream still supports seven species of native south-western fishes.[27] These species are now isolated in Aravaipa Creek as remnants of the former and larger fish fauna of San Pedro River (which was formerly a major tributary to Gila River) which is now a dry river bed at the mouth of Aravaipa Creek (north of Mammoth, Pinal County) except at times of flood or run-off.

The high mountain stream habitat of the vanishing Arizona native trout (*Salmo gilae*) is shown in Figure 64.

The ranch pond is a characteristic aquatic habitat in Arizona and is a focal point for vertebrate wildlife. Figure 63 shows a ranch pond (represo, or dirt tank) made of earth, and with a depth of four to five feet. It was successfully planted with the popular bass-bluegill combination at the time the photograph was taken, November 8, 1952. The locality is at 5,000 feet elevation, on the J. A. Jones Ranch in Parker Canyon, south side of the Huachuca Mountains, Cochise County.

The wide variety of Arizona's sport fishing lakes, from the desert near sea level to coniferous forest at over 9,000 feet elevation, are indicated in Figures 59-62. It is of interest that the most widely fished trout in Arizona, rainbow trout (*Salmo gairdneri*), is stocked and taken in each of these lakes where they are associated with cold-water game species in Big Lake in the White Mountains (*ca.* 9,000 feet, Fig. 61), and with warm-water game species in mid-elevation Peña Blanca Lake in southern Arizona near Nogales, Sonora (*ca.* 4,000 feet, Fig. 62), as well as in low-elevation lake Mohave (*ca.* 700 feet, Fig. 60) and other desert area impoundments in the lower Colorado River near sea level.

---

[27] Longfin dace (*Agosia chrysogaster*), spikedace (*Meda fulgida*), Gila sucker (*Pantosteus clarki*), Sonora sucker (*Catostomus insignis*), loach minnow (*Tiaroga cobitis*), Colorado chub (*Gila robusta*), and speckled dace (*Rhinichthys osculus*).

Fig. 58. The Colorado River in Grand Canyon, between Bright Angel Creek and Lake Mead. The shrubby riparian border of mesquite is a characteristic of the banks of the river in the lower part of the gorge, where the Mohave Desert penetrates upstream well into the canyon. Photo by E. Tad Nichols.

Fig. 59. Aerial view of Iceberg Canyon in upper (eastern) end of Lake Mead, on Arizona-Nevada state line. The nearly barren volcanic hills and mountains in this northeastern part of the Mohave Desert support a few scattered shrubs such as brittlebush (*Encelia farinosa*) and occasional clumps of grass such as tanglehead (*Heteropogon contortus*). Photo by E. Tad Nichols.

Fig. 60. Lake Mohave, on Colorado River in Mohave County, is situated in desert-scrub at 650 ft. elevation (Lower Sonoran). It is one of the deep and narrow reservoir lakes in the lower Colorado River which support rainbow trout "all-year-around in the desert," in addition to bass, catfish, panfish, and other warm water species. Photo by Arizona Game and Fish Department.

Fig. 61. Big Lake, in the White Mountains, Apache County, situated in coniferous forest, *ca.* 9,000 ft. (Boreal). Rainbow trout, Arizona's primary stocked trout species, are successfully planted here, as well as in other high mountain lakes, as fry, fingerlings, and catchable-size fish. Photo by Arizona Game and Fish Department.

Fig. 62. Peña Blanca Lake, Santa Cruz County, 3,800 ft.; a mid-elevation lake in an oak-grass landscape near lower edge of evergreen oak woodland. Upper Sonoran. Since filling in 1958, it has been stocked with largemouth bass, black crappie, channel catfish, catchable-size rainbow trout, and threadfin shad as a forage species. Photo by Arizona Game and Fish Department.

Fig. 63. Ranch Pond, oak woodland. Upper Sonoran. J. A. Jones ranch headquarters, 5,000 ft., Parker Canyon, south side Huachuca Mountains, Cochise County. Planted to bass-bluegill combination. Photo by Velma J. Vance, November 3, 1952.

Fig. 64. East Fork of White River, White Mountains, southwest of Baldy Peak, Apache County, May 17, 1950. Boreal. This is the forest stream habitat of the native Arizona trout, *Salmo gilae* Miller. Photo by Robert Rush Miller.

Fig. 65. Riparian Woodland. Lower Sonoran. Aravaipa Creek at 2,600 ft., tributary to San Pedro River (now a dry bed), paloverde-sahuaro foothills, west side Galiuro Mountains, Pinal County. This spring-fed desert stream has a permanent flow of 2-3 feet per second and supports seven species of native fishes (see text). The riparian cottonwood-willow gallery association contains other broadleaf winter-deciduous trees such as sycamore, ash, and walnut.

Fig. 66. Riparian Woodland. Lower Sonoran. Aravaipa Creek at same locality as Figure 65. The *non-riparian climax* of the surrounding desert is typical paloverde-sahuaro on foothills in the Arizona upland section of the Sonoran Desert. The broadleaf trees are the canopy counterpart of a distinctive *climax riparian* community confined here to the desert drainageway.

LITERATURE CITED

Adams, C. C., G. P. Burns, T. L. Hankinson, B. Moore, and N. Taylor

1920a Plants and animals of Mount Marcy, New York, Part I. Ecology, 1:71-94.

1920b Plants and animals of Mount Marcy, New York, Part III. Ecology, 1:274-288.

Aldous, A. E., and H. L. Shantz

1924 Types of vegetation in the semiarid portion of the United States and their economic significance. J. Agr. Research, 28:99-128.

Aldrich, J. W. and H. Friedmann

1943 A revision of the ruffed grouse. Condor, 45:85-103.

Alexander, W. H.

1935 The distribution of thunderstorms in the United States. Monthly Weather Rev., 63:157-158.

Allee, W. C.

1926 Some interesting animal communities of northern Utah. Sci. Monthly, 23:481-495.

Allee, W. C., A. E. Emerson, O. Park, T. Park, and K. P. Schmidt

1949 Principles of animal ecology. W. B. Saunders Co., Philadelphia.

Allen, J. A.

1871 On the mammals and winter birds of east Florida with an examination of certain assumed specific characters in birds and a sketch of the bird faunae of Eastern North America. Bull. Mus. Comp. Zool., Harvard College, 2:161-450.

1892 The geographic distribution of North American mammals. Bull. Amer. Mus. Nat. Hist., 4:199-243.

1893 The geographical origin and distribution of North American birds, considered in relation to faunal areas of North America. Auk, 10:97-150.

American Ornithologists' Union Committee

1910 Check list of North American birds. Third edition, New York.

Arnberger, L. P.

1947 Flowering plants and ferns of Walnut Canyon. Plateau, 20:29-36.

1952 Flowers of the Southwest mountains, Southwest Monuments Assoc., Pop. Series, 7:1-112.

Arnold, J. F.

1950 Changes in ponderosa pine bunchgrass ranges in northern Arizona resulting from pine regeneration and grazing. J. Forestry, 48:118-126.

Arnold, J. F. and W. L. Schroeder

1955 Juniper control increases forage production on the Fort Apache Indian Reservation. U. S. Dep. Agr., Rocky Mountain Forest and Range Exp. Sta., Sta. Paper 18:1-35.

Axelrod, D. I.

1950 Evolution of desert vegetation in Western North America. *In* D. I. Axelrod: Studies in late Tertiary paleobotany. Carnegie Inst. Washington, Publ. 590:1-323.

1958 Evolution of the Madro-Tertiary Geoflora. Botan. Rev., 24:433-509.

Bailey, F. M.

1902 Handbook of birds of the western United States. Houghton Mifflin, Boston, 514 p.

1928 Birds of New Mexico. New Mexico Dept. Game and Fish, Santa Fe.

1939 Among the birds in Grand Canyon Country. U. S. Dept. Interior, Nat. Park Serv., Washington, 211 p.

Bailey, V.

1902 Biological survey of Texas. U. S. Dept. Agr., N. Amer. Fauna, 25:1-222.

1913 Life-zones and crop-zones of New Mexico. U. S. Dept. Agr., N. Amer. Fauna, 35:1-100.

1926 A biological survey of North Dakota. U. S. Dept. Agr., N. Amer. Fauna, 49:1-229.

1929 Life Zones of the Grand Canyon. Grand Canyon Nat. Notes, 3:2-5.

1931 Mammals of New Mexico. U. S. Dept. Agr., N. Amer. Fauna, 53:1-412.

1935 Mammals of the Grand Canyon region. Grand Canyon Nat. Hist. Assoc., Nat. Hist. Bull., 1:1-42.

1936 The mammals and life zones of Oregon. U. S. Dept. Agr., N. Amer. Fauna, 55:1-416.

Baker, F. S.

1945 Forest cover types of western North America. Pamphlet, Soc. Amer. Foresters, p. 1-35.

Baker, W. H.

1956 Plants of Iron Mountain, Rogue River Range, Oregon. Amer. Midl. Nat., 56:1-53.

Ball, E. D., E. R. Tinkham, R. Flock, and C. T. Vorhies

1942 The grasshoppers and other orthoptera of Arizona. Univ. Ariz. Agr. Exp. Sta. Tech. Bull., 93:1-373.

Bancroft, G.

1926 The faunal areas of Baja California del Norte. Condor, 28:209-215.

Bates, C. G.

1924 Forest types in the central Rocky Mountains as affected by climate and soil. U. S. Dep. Agr. Bull., 1233:1-152.

Beaman, J. H.

1962 The timberlines of Iztaccihuatl and Popocatepetl, Mexico. Ecology, 43:377-385.

Beaufort, L. F. de

1951 Zoogeography of the land and inland waters. Sidgwick and Jackson, London.

Benson, L.

1940 The cacti of Arizona. Univ. Ariz. Press, Tucson.

Benson, L. and R. A. Darrow

1944 A manual of southwestern desert trees and shrubs. Univ. Ariz. Biol. Sci. Bull., 6:1-411.

1954 The trees and shrubs of the southwestern deserts. 2nd ed. of Benson and Darrow (1944). Univ. Ariz. Press, Tucson.

Billings, W. D. and L. C. Bliss

1959 An alpine snowbank environment and its effect on vegetation, plant development and productivity. Ecology, 40:388-397.

Blair, W. F.

1940 A contribution to the ecology and faunal relationships of the mammals of the Davis Mountain Region, Southwestern Texas. Univ. Mich. Mus. Zool., Misc. Publ., 46:1-39.

1950 The biotic provinces of Texas. Texas J. Sci., 2:93-117.

1952 Mammals of the Tamaulipan biotic province in Texas. Texas J. Sci., 4:230-250.

1961 Vertebrate speciation. (Editor). Univ. Texas Press, Austin.

Blair, W. F. and T. H. Hubbell

1938 The biotic districts of Oklahoma. Amer. Midl. Nat., 20:425-454.

Blumer, J. C.

1909 On the plant geography of the Chiricahua Mountains. Science, 30: 720-724.

1910 A comparison between two mountain sides. Plant World, 13:134-140.

1911 Change of aspect with altitude. Plant World, 14:236-248.

1912 Notes on the phytogeography of the Arizona desert. Plant World, 15: 183-189.

Brandegee, T. S.

1880 Timberline in the Wasatch Range. Bot. Gaz., 5:125-126.

Brandt, H.

1951 Arizona and its bird life. The Bird Research Foundation, Cleveland, Ohio.

Branscomb, B. L.

1958 Shrub invasion of a southern New Mexico desert grassland range. J. Range Mgmt., 11:129-132.

Bray, W. L.

1905 The vegetation of the sotol country in Texas. Univ. Texas, Bull. 60, Sci. Ser., 6:1-24.

Britton, N. L.

1889 A list of plants collected at Fort Verde and vicinity and in the Mogollon and San Francisco Mountains, Arizona, 1884-1888, by Dr. E. A. Mearns, U.S.A. Trans. New York Acad. Sci., 8:61-76.

Brown, A. E.

1903 The faunal relations of Texas reptiles. Proc. Acad. Nat. Sci., Phila., 55:551-558.

Brown, A. L.

1950 Shrub invasion of southern Arizona desert grassland. J. Range Mgmt., 3:172-177.

Brown, R. J. and J. T. Curtis

1952 The upland conifer-hardwood forests of northern Wisconsin. Ecol. Monog., 22:217-234.

Bryan, K.

1925 The Papago Country, Arizona. A geographic, geologic and hydrologic reconnaissance with a guide to desert watering places. U. S. Geol. Surv. Water Supply Paper 449:1-436.

1928 Change in plant associations by change in ground water level. Ecology, 9:474-478.

Burt, W. H.

1938 Faunal relationships and geographic distribution of mammals in Sonora, Mexico. Univ. Mich. Mus. Zool., Misc. Publ., 39:1-77.

Cain, S. A.
1939 The climax and its complexities. Amer. Midl. Nat., 21:146-181.
1944 Foundations of plant geography. Harper and Brothers, New York.
1950 Life-forms and phytoclimate. Bot. Rev., 16:1-32.

Cain, S. A. and G. M. de Oliviera Castro
1959 Manual of vegetation analysis. Harper and Brothers, New York.

Campbell, R. S. and E. H. Bomberger
1934 The occurrence of *Gutierrezia sarothrae* on *Bouteloua eriopoda* ranges in Southern New Mexico. Ecology, 15:49-61.

Cannon, W. A.
1916 Distribution of the cacti with especial reference to the role played by the root response to soil temperature and soil moisture. Amer. Nat., 50:435-442.

Cary, M.
1911 A biological survey of Colorado. U. S. Dept. Agr., N. Amer. Fauna, 33:1-256.
1917 Life zone investigations in Wyoming. U. S. Dept. Agr., N. Amer. Fauna, 42:1-95.

Castetter, E. F.
1956 The vegetation of New Mexico. New Mexico Quarterly, 26:257-288.

Chaney, R. W., C. Condit, and D. I. Axelrod
1944 Pliocene floras of California and Oregon. Carnegie Inst. Washington, Publ. 553:1-407.

Clark, H. W.
1937 Association types in the north Coast Ranges of California. Ecology, 18:214-230.

Clarke, G. L.
1954 Elements of ecology. John Wiley and Sons, New York.

Clements, F. E.
1920 Plant indicators. Carnegie Inst. Washington, Publ. 290:1-388.

Clements, F. E. and V. E. Shelford
1939 Bio-ecology. John Wiley and Sons, New York.

Cockerell, T. D. A.
1897 Life zones in New Mexico, Pt. I. New Mex. Agr. Exp. Stat. Bull., 24:1-44.
1898 Life zones in New Mexico, Pt. II. New Mex. Agr. Exp. Sta. Bull., 28:137-179.
1900 The lower and middle Sonoran Zones in Arizona and New Mexico. Amer. Nat., 34:285-293.
1941 Observations on plants and animals in northwestern Baja California, Mexico, with descriptions of new bees. Trans. San Diego Soc. Nat. Hist., 9:337-352.

Cockrum, E. L.
1955 Manual of mammalogy. Burgess Publ. Co., Minneapolis.
1963 An annotated check list of the mammals of Arizona. *In* The vertebrates of Arizona. Univ. Ariz. Press, Tucson.

Coleman, B. B., W. C. Muenscher and D. R. Charles
1956 A distributional study of the epiphytic plants of the Olympic Peninsula, Washington. Amer. Midl. Nat., 56:54-87.

Cooke, W. B.

1941 The problem of life-zones on Mount Shasta, California. Madroño, 6:49-55.

Cooper, C. F.

1960 Changes in vegetation structure and growth of southwestern pine forests since white settlement. Ecol. Monog., 30:129-164.

1961a Controlled burning and watershed condition in the White Mountains of Arizona. J. Forestry, 59:438-442.

1961b Pattern in ponderosa pine forests. Ecology, 42:493-499.

Cope, E. D.

1866 On the reptilia and batrachia of the Sonoran Provinces of the Nearctic Region. Proc. Acad. Nat. Sci. Phila., 18:300-314.

1875 On the geographical distribution of the vertebrata of the Regnum Nearcticum, with especial reference to the batrachia and reptilia. Bull. U. S. Nat. Mus., 1:55-95.

Cox, C. F.

1933 Alpine plant succession on James Peak, Colorado. Ecol. Monog., 3: 299-372.

Croft, A. R.

1933 Notes on pinyon-juniper reforestation. Grand Canyon Nat. Hist. Notes, 8:151-154.

Cronemiller, F. P.

1942 Chaparral. Madroño, 6:199.

Curtis, J. T. and R. P. McIntosh

1951 An upland forest continuum in the prairie-forest border region of Wisconsin. Ecology, 32:476-496.

Dalquest, W. W.

1948 Mammals of Washington. Univ. Kans. Mus. Nat. Hist., Publ., 2:1-444.

Dansereau, P.

1957 Biogeography, an ecological perspective. The Ronald Press Co., New York.

Darlington, P. J. Jr.

1957 Zoogeography: The geographical distribution of animals. John Wiley and Sons, Inc., New York.

Darrow, R. A.

1944 Arizona range resources and their utilization. I. Cochise County. Univ. Ariz. Agr. Exp. Sta. Tech. Bull., 103:311-366.

1961 Origin and development of the vegetational communities of the Southwest. *In* Bioecology of the arid and semiarid lands of the Southwest. New Mexico Highlands Univ., Bull. 212:30-47.

Darwin, C.

1859 The origin of species by means of natural selection. First Modern Library Edition (1936), New York.

Daubenmire, R. F.

1938 Merriam's life zones of North America. Quart. Rev. Biol., 13:327-332.

1943a Soil temperature versus drouth as a factor determining lower altitudinal limits of trees in the Rocky Mountains. Botan. Gaz., 105:1-13.

1943b Vegetational zonation in the Rocky Mountains. Botan. Rev., 9:325-393.

1954 Alpine timberlines in the Americas and their interpretation. Butler Univ. Botan. Studies, 11:119-136.

1959 Plants and environment. A textbook of plant autecology. 2nd ed. John Wiley and Sons, Inc., New York.

Davis, W. B.

1939 The recent mammals of Idaho. Caxton Printers, Caldwell, Idaho.

Deaver, C. F. and H. S. Haskell

1955 Ferns and flowering plants of Havasu Canyon. Plateau, 28:11-23.

Deevey, E.

1949 Biogeography of the Pleistocene. Part I. Europe and North America. Bull. Geol. Soc. Amer., 60:1315-1416.

Dellenbaugh, F. S.

1932 The Painted Desert. Science, 76:437.

Denyes, H. A.

1956 Natural terrestrial communities of Brewster County, Texas, with special reference to the distribution of the mammals. Amer. Midl. Nat., 55: 289-320.

Dessauer, H. C., W. Fox, and F. H. Pough

1962 Starch-gel electrophoresis of transferrins, esterases, and other plasma proteins of hybrids between two subspecies of whiptailed lizard (Genus *Cnemidophorus*). Copeia, 1962 (4):767-774.

Detling, L. E.

1948 Concentration of environmental extremes as the basis for vegetational areas. Madroño, 9:169-185.

Dice, L. R.

1922 Biotic areas and ecological habitats as units for the statement of animal and plant distribution. Science, 55:335-338.

1939 The Sonoran Biotic Province. Ecology, 20:118-129.

1943 The biotic provinces of North America. Univ. Mich. Press, Ann Arbor.

1952 Natural Communities. Univ. Mich. Press, Ann Arbor.

Dice, L. R. and P. M. Blossom

1937 Studies of mammalian ecology in southwestern North America with special attention to the colors of desert mammals. Carnegie Inst. Washington, Publ. 485:1-129.

Dickerman, R. W.

1954 An ecological survey of the Three-Bar Game Management Unit located near Roosevelt, Arizona. Master's Thesis, University of Arizona, Tucson.

Ditmer, H. J.

1951 Vegetation of the Southwest — past and present. Texas J. Sci., 3:350-355.

Dodge, N. N.

1936 Trees of Grand Canyon National Park. Grand Canyon National Hist. Assoc., Nat. Hist. Bull., 3:1-69.

1938 Amphibians and reptiles of Grand Canyon National Park. Grand Canyon Nat. Hist. Assoc., Nat. Hist. Bull., 9:1-55.

1958 Flowers of the Southwest deserts. Southwestern Monuments Assoc., Pop. Ser., 4:1-112.

Dorf, E.

1960 Climatic changes of the past and present. Amer. Scientist, 48:341-364.

Dorroh, J. H., Jr.

1946 Certain hydrologic and climatic characteristics of the Southwest. Univ. New Mex., Publ. Eng., 1:1-64.

Durrant, S. D.

1952 Mammals of Utah, taxonomy and distribution. Univ. Kans. Mus. Nat. Hist. Publ., 6:1-549.

1959 The nature of mammalian species. *In* Species: Modern Concepts. J. Ariz. Acad. Sci., 1:18-21.

Eastwood, A.

1919 Early spring at the Grand Canyon near El Tovar. Plant World, 22:95-99.

Egler, F. E.

1951 A commentary on American plant ecology, based on the textbooks of 1947-1949. Ecology, 32:673-695.

Fenneman, N. M.

1931 Physiography of the western United States. McGraw-Hill Book Co., New York.

Findley, J. S. and S. Anderson

1956 Zoogeography of the montane mammals of Colorado. J. Mammal., 37: 80-82.

Forsaithe, C. C.

1920 Anatomical reduction in some alpine plants. Ecology, 1:124-135.

Fosberg, F. R.

1938 The Lower Sonoran in Utah. Science, 87:39-40.

1959 Upper limits of vegetation of Mauna Loa, Hawaii. Ecology, 40:144-146.

1961 Classification of vegetation for general purposes. Trop. Ecol., 1:1-28.

Gardner, J. L.

1951 Vegetation of the creosotebush area of the Rio Grande Valley in New Mexico. Ecol. Monog., 21:379-403.

Garth, J. S.

1935 Butterflies of Yosemite National Park. Bull. So. Calif. Acad. Sci., 34: 37-75.

1950 Butterflies of Grand Canyon National Park. Grand Canyon Nat. Hist. Assoc., Bull. 11:1-52.

Gentry, H. S.

1942 Rio Mayo plants. A study of the flora and vegetation of the valley of the Rio Mayo, Sonora. Carnegie Inst. Washington, Publ. 527:1-316.

Gleason, H. A.

1917 The structure and development of the plant association. Bull. Torrey Botan. Club, 44:463-481.

1926 The individualistic concept of the plant association. Bull. Torrey Botan. Club, 53:7-26.

Glendening, G. E.

1952 Some quantitative data on the increase of mesquite and cactus on a desert grassland range in southern Arizona. Ecology, 33:319-328.

Gloyd, H. K.

1932 A consideration of the faunal areas of southern Arizona based on the distribution of amphibians and reptiles. Anat. Record, 54:109-110 (abstract).

1937 A herpetological consideration of faunal areas in southern Arizona. Bull. Chicago Acad. Sci., 5:79-136.

Goldman, E. A. and R. T. Moore

1945 The biotic provinces of Mexico. J. Mammal., 26:347-360.

Good, R.
1947 The geography of the flowering plants. Longmans, Green and Co., London.

Gould, F. W.
1951 Grasses of southwestern United States. Univ. Ariz. Biol. Sci. Bull., 7:1-343.

Graham, E. H.
1937 Botanical studies in the Uinta Basin of Utah and Colorado. Ann. Carnegie Mus., 26:1-432.

Grater, R. K.
1937 Check-list of birds of Grand Canyon National Park. Grand Canyon Nat. Hist. Assoc., Nat. Hist. Bull., 8:1-55.

Gray, J.
1961 Early Pleistocene paleoclimatic record from the Sonoran Desert, Arizona. Science, 113:38-39.

Green, C. R.
1962 Arizona climate—Supplement No. 1. Probabilities of temperature occurrence in Arizona and New Mexico. Univ. Ariz. Press, Tucson.

Griggs, R. F.
1934 The edge of the forest in Alaska and the reasons for its position. Ecology, 15:80-96.
1938 Timberlines in the northern Rocky Mountains. Ecology, 19:548-564.
1946 The timberlines of North America and their interpretation. Ecology, 27:275-289.

Grinnell, J.
1908 The biota of the San Bernardino Mountains. Univ. Calif. Publ. Zool., 5:1-170.
1914 An account of the mammals and birds of the Lower Colorado Valley, with especial reference to the distributional problems presented. Univ. Calif. Publ. Zool., 12:1-217.
1915 A distributional list of the birds of California. Pacific Coast Avifauna, 11:1-217.
1928 A distributional summation of the ornithology of Lower California. Univ. Calif. Publ. Zool., 32:1-300.
1935 A revised life-zone map of California. Univ. Calif. Publ. Zool., 40: 327-330.

Grinnell, J., J. Dixon, and J. M. Linsdale
1930 Vertebrate natural history of the Lassen Peak region. Univ. Calif. Publ. Zool., 35:1-594.

Grinnell, J. and T. I. Storer
1924 Animal life in Yosemite. Univ. Calif. Press, Berkeley.

Grinnell, J. and H. S. Swarth
1913 An account of the birds and mammals of the San Jacinto area of southern California, with remarks upon the behavior of geographic races on the margins of their habitats. Univ. Calif. Publ. Zool., 10:197-406.

Grisebach, A.
1838 Ueber den Einfluss des Climas auf die Begränzung der natürlichen Floren. Linnaea, 12:159-200.

Hall, E. R.
1946 Mammals of Nevada. Univ. Calif. Press, Berkeley.

Hall, H. M. and J. Grinnell
1919 Life-zone indicators in California. Proc. Calif. Acad. Sci., ser. 4, 9:37-67.

Hanson, H. C.
1924 A study of the vegetation of northeastern Arizona. Univ. Nebr. Studies, 24:85-175.

Hargrave, L. L.
1933a Bird life of the San Francisco Mountains, Arizona. Number One: Summer Birds. Mus. N. Ariz., Mus. Notes, 5:57-60.
1933b Bird Life of the San Francisco Mountains, Arizona. Number Two: Winter Birds. Mus. N. Ariz., Mus. Notes, 6:27-34.
1936 Bird life of the San Francisco Mountains, Arizona. Number Three: Land birds known to nest in the pine belt. Mus. N. Ariz., Mus. Notes, 9:47-50.

Harshberger, J. W.
1911 Phytogeographic survey of North America. G. E. Stechert, New York.

Hart, F. C.
1937 Precipitation and run-off in relation to altitude in the Rocky Mountain region. J. Forestry, 35:1005-1010.

Haskell, H. S.
1958 Flowering plants in Glenn Canyon. Late summer aspect. Plateau, 31:1-3.

Haskell, H. S., and C. F. Deaver
1955 Plant life-zones. Plateau, 27:21-24.

Hastings, J. R.
1959 Vegetation change and arroyo cutting in southeastern Arizona. J. Ariz. Acad. Sci., 1:60-67.

Haviland, M. D.
1926 Forest, steppe, and tundra. Cambridge Univ. Press, London.

Heald, W. F.
1951 Sky islands of Arizona. Nat. Hist., 60:56-63, 95-96.

Hesse, R., W. C. Allee, and K. P. Schmidt
1937 Ecological animal geography. John Wiley and Sons, New York.

Hoffman, W. J.
1877 The distribution of vegetation in portions of Nevada and Arizona. Amer. Nat., 11:336-343.

Hoffmeister, D. F.
1955 Mammals new to Grand Canyon National Park, Arizona. Plateau, 28:1-7.

Hoffmeister, D. F., and W. W. Goodpaster
1954 The mammals of the Huachuca Mountains, Southeastern Arizona. Ill. Biol. Monog., 24:1-152. Univ. Ill. Press, Urbana.

Holdenreid, R. and H. B. Morlan
1956 A field study of wild mammals and fleas of Santa Fe County, New Mexico. Amer. Midl. Nat., 55:369-381.

Holm, T.
1927 The vegetation of the alpine region of the Rocky Mountains in Colorado. Mem. U. S. Nat. Acad. Sci., 19:1-45.

Holzman, B.
1937 Sources of moisture for precipitation in the United States. U. S. Dept. Agr. Tech. Bull. 589:1-41.

Hopkins, A. D.

1918 Periodical events and natural laws as guides to agricultural research and practice. Monthly Weather Rev., Suppl. 9:1-42.

1938 Bioclimatics. A science of life and climatic relations. U. S. Dept. Agr., Misc. Publ. 280:1-188.

Howell, A. B.

1917 Birds of the islands of the coast of Southern California. Pac. Coast Avifauna, 12:1-127.

1923 Influences of the Southwestern deserts upon the avifauna of California. Auk, 40:584-592.

Howell, A. H.

1921 A biological survey of Alabama: I. Physiography and life-zones: II. Mammals. U. S. Dept. Agr., N. Amer. Fauna, 45:1-88.

1938 Revision of the North American ground squirrels, with a classification of the North American Sciuridae. U. S. Dept. Agr., N. Amer. Fauna, 56:1-226.

Howell, J., Jr.

1941 Piñon and juniper woodlands of the Southwest. J. Forestry, 39:542-545.

Huey, L. M.

1942 A vertebrate faunal survey of the Organ Pipe Cactus National Monument, Arizona. Trans. San Diego Soc. Nat. Hist., 9:355-375.

Humboldt, A.

1807 Essai sur la geographie des plantes. Paris (1805).

Humphrey, R. R.

1949 Fire as a means of controlling velvet mesquite, burroweed and cholla on southern Arizona ranges. J. Range Mgmt., 2:173-182.

1950 Arizona range resources. II. Yavapai County. Univ. Ariz. Agr. Exp. Sta. Bull., 229:1-55.

1953a Forage production on Arizona ranges. III. Mohave County. Univ. Ariz. Agr. Exp. Sta. Bull., 244:1-79.

1953b The desert grassland, past and present. J. Range Mgmt., 6:159-164.

1955 Forage production on Arizona ranges. IV. Coconino, Navajo, Apache Counties. A study in range condition. Univ. Ariz. Agr. Exp. Sta. Bull., 266:1-84.

1958 The desert grassland. A history of vegetational change and an analysis of causes. Botan. Rev., 24:193-252.

1960 Forage production on Arizona ranges. V. Pima, Pinal, and Santa Cruz Counties. Univ. Ariz. Agr. Exp. Sta. Bull., 302:1-138.

1962 Range ecology. Ronald Press Co., New York.

Humphrey, R. R. and A. C. Everson

1951 Effect of fire on a mixed grass-shrub range in southern Arizona. J. Range Mgmt., 4:264-266.

Humphrey, R. R., A. L. Brown, and A. C. Everson

1956 Common Arizona range grasses. Univ. Ariz. Agr. Exp. Sta. Bull., 243:1-102.

Humphrey, R. R. and L. A. Mehrhoff

1958 Vegetation changes on a southern Arizona grassland range. Ecology, 39:720-726.

Ingles, L. G.

1947 Mammals of California. Stanford Univ. Press, Stanford.

Ives, J. C.

1861 Report upon the Colorado River of the West. U. S. Senate Ex. Doc., 36th Congress, 1st Session, pt. 1, p. 1-139.

Ives, R. L.

1949 Climate of the Sonoran Desert. Ann. Assoc. Amer. Geog., 34:143-187.

Jameson, D. A., J. A. Williams, and E. W. Wilton

1962 Vegetation and soils of Fishtail Mesa, Arizona. Ecology, 43:403-410.

Jameson, D. L. and A. G. Flury

1949 The reptiles and amphibians of the Sierra Vieja Range of southwestern Texas. Tex. J. Sci., 1:54-79.

Jenks, Randolph

1931 Ornithology of the life-zones. Summit of San Francisco Mountains to bottom of Grand Canyon. Grand Canyon Nat. Park, Tech. Bull., 5:1-31.

Jepson, W. L.

1925 A manual of the flowering plants of California. Univ. Calif. Assoc. Stud. Store, Berkeley.

Johnson, D. H., M. D Bryant, and A. H. Miller

1948 Vertebrate animals of the Providence Mountains area of California. Univ. Calif. Publ. Zool., 48:221-376.

Johnson, I. M.

1919 The flora of the pine belt of the San Antonio Mountains of Southern California. Plant World, 22:71-90, 105-121.

Jones, G. N.

1936 A botanical survey of the Olympic Peninsula, Washington. Univ. Wash. Publ. Biol., 5:1-286.

1938 The flowering plants and ferns of Mount Rainier. Univ. Wash. Publ. Biol., 7:1-142.

Jurwitz, L. R.

1953 Arizona's two-season rainfall pattern. Weatherwise, 6:96-99.

Kearney, T. H. and R. H. Peebles

1942 Flowering plants and ferns of Arizona. U. S. Dept. Agr., Misc. Publ., 423:1-1069.

1951 Arizona Flora. Univ. Calif. Press, Berkeley.

1960 Arizona Flora. 2nd ed. with suppl. by J. T. Howell, Elizabeth McClintock and collaborators. Univ. Calif. Press, Berkeley.

Kendeigh, S. C.

1932 A study of Merriam's temperature laws. Wilson Bull., 44:129-143.

1948 Bird populations and biotic communities in northern lower Michigan. Ecology, 29:101-114.

1954 History and evolution of various concepts of plant and animal communities in North America. Ecology, 35:152-171.

1961 Animal ecology. Prentice-Hall, Inc., Engelwood Cliffs.

Keppel, R. V., J. E. Fletcher, J. L. Gardner, and K. G. Renard

1958 Southwest Watershed Hydrology Studies Group, Tucson. Annual Progress Report, 1958-1960.

Kincer, J. B.

1922 Precipitation and humidity. *In* Atlas of American Agriculture. U. S. Printing Office, Washington, D. C.

1941 Climate and man. U. S. Dept. Agr., Yearbook of Agr.

1946 Our Changing climate. Trans. Amer. Geophys. Union, 27:342-347.

Law, J. E.
1929 A discussion of faunal influences in southern Arizona. Condor, 31: 216-220.

Le Conte, J. L.
1859 The coleoptera of Kansas and eastern New Mexico. Smithsonian Contrib. Knowl., 11:1-58.

Leiberg, J. B., T. F Rixon, and A. Dowell
1904 Forest conditions in the San Francisco Mountains Forest Reserve, Arizona. U. S. Geol. Survey, Prof. Paper 22 (Ser. H, Forestry, 7) :1-95.

Leopold, A.
1924 Grass, brush, timber, and fire. J. Forestry, 22:1-10.

Leopold, A. S.
1950 Vegetation zones of Mexico. Ecology, 31:507-518.

Lewis, H.
1959 The nature of plant species. *In* Species: Modern Concepts. J. Ariz. Acad. Sci., 1:3-7.

Lindsey, A. A.
1951 Vegetation and habitats in a southwestern volcanic area. Ecol. Monog., 21:277-253.

Little, E. L.
1941 Alpine flora of San Francisco Mountains, Arizona. Madroño, 6:65-81.
1950 Southwestern trees; a guide to the native species of New Mexico and Arizona. U. S. Dept. Agr., Agri. Handbook 9:1-109.
1953 Check list of native and naturalized trees of the United States (including Alaska). U. S. Dept. Agr., Handbook 41:1-472.

Livingston, B. E. and F. Shreve
1921 The distribution of vegetation in the United States as related to climatic conditions. Carnegie Inst. Washington, Publ. 284: 1-590.

Lloyd F. E.
1907 Pima Canyon and Castle Rock in the Santa Catalina Mountains. Plant World, 10:251-259.

Loew, O.
1875 Report upon mineralogical, agricultural, and chemical conditions observed in portions of Colorado, New Mexico, and Arizona in 1873. U. S. Geog. Surv. W. 100th Merid. (Wheeler), Geol. III, pt. 6, p. 569-661.

Lowe, C. H.
1955 The eastern limit of the Sonoran Desert in the United States with additions to the known herpetofauna of New Mexico. Ecology, 36: 343-345.
1959a Contemporary biota of the Sonoran Desert: Problems. *In* University of Arizona, Arid Lands Colloquia, 1958-59:54-74.
1959b Summation—specific and infraspecific variation. *In* Species: Modern Concepts. J. Ariz. Acad. Sci., 1:31-34.
1961 Biotic communities in the sub-Mogollon region of the Inland Southwest. J. Ariz. Acad. Sci., 2:40-49.
1963 An annotated check list of the amphibians and reptiles of Arizona. *In* the Vertebrates of Arizona. Univ. Ariz. Press, Tucson.

Lull, H. W. and L. Ellison
1950 Precipitation in relation to altitude in central Utah. Ecology, 31:479-484.

Lutz, F. E.

1934 From low to high. Grand Canyon Nat. Notes, 9:327-329.

MacDougal, D. T.

1908a Across Papagueria. Bull. Amer. Geog. Soc., 40:1-21.

1908b Botanical features of North American deserts. Carnegie Inst. Washington, Publ. 99:1-111.

1921 The reactions of plants to new habitats. Ecology, 2:1-20.

Marks, J. B.

1950 Vegetation and soil relations in the lower Colorado desert. Ecology, 31:176-193.

Marshall, J. T., Jr.

1956 Summer birds on the Rincon Mountains, Saguaro National Monument, Arizona. Condor, 58:81-97.

1957 Birds of the pine-oak woodland in southern Arizona and adjacent Mexico. Pacific Coast Avifauna, 32:1-125.

1962 Land use and native birds of Arizona. J. Ariz. Acad. Sci., 2:75-77.

Martin, P. S.

1959 Terrestrial communities in the Pleistocene. *In* Problems of the Pleistocene epoch and Arctic area. McGill Univ. Mus. Publ. 1:26-38.

1961 Southwestern animal communities in the late Pleistocene. New Mexico Highlands Univ. Bull., Catalog Issue, 1961:56-66.

Martin, P. S. and J. Gray

1962 Pollen analysis and the Cenozoic. Science, 137:103-111.

Martin, P. S., J. Schoenwetter, and B. C. Arms

1961 Southwestern palynology and prehistory. The last 10,000 years. Geochronology Lab., Univ. Ariz. Press, Tucson.

Martin, W. P. and J. E. Fletcher

1943 Vertical zonation of great soil groups on Mt. Graham, Arizona, as correlated with climate, vegetation, and profile characteristics. Univ. Ariz. Agr. Exp. Sta. Tech. Bull. 99:89-153.

Martinez, M.

1945 Las pineaceas mexicanas. Instituto de Biología, México, D.F.

1948 Los pinos mexicanos. Botas, México, D.F.

Maslin, T. P.

1959 The nature of amphibian and reptilian species. *In* Species: Modern Concepts. J. Ariz. Acad. Sci., 1:8-17.

Mason, H. L.

1947 Evolution of certain floristic associations in western North America. Ecol. Monog., 17:203-210.

Mayr, E., E. G. Linsley and R. L. Usinger

1953 Methods and principles of systematic zoology. McGraw-Hill Book Co., New York.

McDonald, J. E.

1956 Variability of precipitation in an arid region; a survey of characteristics for Arizona. Univ. Ariz. Inst. Atmos. Physics, Tech. Rept., 1:1-88.

McDougall, W. B.

1957 Botany of the Museum and Colton Ranch Area. I. General Ecology. Plateau, 29:81-87.

McGee, W. J. and W. D. Johnson
1896 Seriland. Nat. Geog., 7:125-133.

McHenry, D. E.
1932a Demonstration life-zone gardens. Grand Canyon Nat. Notes, 7:81-85.
1932b Quaking aspen at Grand Canyon. Grand Canyon Nat. Notes, 7:87.
1933 Woodland parks on the North Rim. Grand Canyon Nat. Notes, 8: 195-198.
1934 Canadian zone plants on the South Rim of Grand Canyon. Grand Canyon Nat. Notes, 9:301-302.
1935 Quaking aspen—its future in the Park. Grand Canyon Nat. Notes, 9: 361-364.

McKee, B. H.
1933 The naming of the Grand Canyon. Grand Canyon Nat. Notes, 8:210-212.

McKee, E. D.
1934 Flora of Grand Canyon National Monument. Grand Canyon Nat. Notes, 9:310-321.
1941 Distribution of the tassel-eared squirrels. Plateau, 14:12-20.

Mead, P.
1930 A brief ecological comparison of life-zones on the Kaibab Plateau. Grand Canyon Nat. Notes, 5:13-17.

Mearns, E. A.
1896 Preliminary diagnoses of new mammals from the Mexican border of the United States. Proc. U. S. Nat. Mus., 19:137-140.
1907 Mammals of the Mexican boundary of the United States. A descriptive catalogue of the species of mammals occurring in that region; with a general summary of the natural history, and a list of trees. Bull. U. S. Nat. Mus., 56:1-530.

Melton, M. A.
1960 Origin of the drainage and geomorphic history of southeastern Arizona. *In* University of Arizona, Arid Lands Colloquia, 1960-61:8-16.

Merkle, J.
1952 An analysis of a pinyon-juniper community at Grand Canyon, Arizona. Ecology, 33:375-384.

Merriam, C. H.
1890 Results of a biological survey of the San Francisco Mountains region and desert of the Little Colorado in Arizona. U. S. Dept. Agr., N. Amer. Fauna, 3:1-136.
1891 Results of a biological reconnaissance of south-central Idaho. U. S. Dept. Agr., N. Amer. Fauna, 5:1-127.
1892 The geographical distribution of life in North America with special reference to the mammalia. Proc. Biol. Soc. Wash., 7:1-64.
1894 Laws of temperature control of the geographic distribution of terrestrial animals and plants. Nat. Geog. Mag., 6:229-238.
1898 Life-zones and crop-zones of the United States. U. S. Dept. Agr., Div. Biol. Surv., Bull. 10:1-79.
1899a Biological Survey of Mount Shasta. U. S. Dept. Agr., N. Amer. Fauna, 16:1-180.
1899b Zone temperatures. Science, 9:116.

Merriam, C. H., V. Bailey, E. W. Nelson, and E. A. Preble
1910 Zone map of North America. U. S. Dept. Agr., Div. Biol. Surv., Washington, D. C. [This map was used as frontispiece for the third A.O.U. check list of North American Birds, 1910.]

Miller, A. H.

1937 Biotic associations and life zones in relation to the Pleistocene birds of California. Condor, 39:248-252.

1946 Vertebrate inhabitants of the piñon association in the Death Valley region. Ecology, 27:54-60.

1951 An analysis of the distribution of the birds of California. Univ. Calif. Publ. Zool., 50:531-643.

Miller, F. H.

1921 Reclamation of grasslands by Utah juniper on the Tusayan National Forest, Arizona. J. Forestry, 19:647-651.

Miller, G. S.

1895 The long-tailed shrews of the eastern United States. U. S. Dept. Agr., N. Amer. Fauna, 10:35-56.

1897 Revision of the North American bats of the family Vespertilionidae. U. S. Dept. Agr., N. Amer. Fauna, 13:1-135.

Miller, R. R.

1954 A drainage map of Arizona. Syst. Zool., 3:80-81.

Miller, R. R. and C. H. Lowe

1963 An annotated check list of the fishes of Arizona. *In* Vertebrates of Arizona. Univ. Ariz. Press, Tucson.

Monson, G.

1942 Notes on some birds of southeastern Arizona. Condor, 44:222-225.

Monson, G. and A. R. Phillips

1941 Bird records from southern and western Arizona. Condor, 43:108-112.

Moore, R. T.

1945 The transverse volcanic biotic province of central Mexico and its relationship to adjacent provinces. Trans. San Diego Soc. Nat. Hist., 10: 217-235.

Muesebeck, C. F. W. and K. V. Krombein

1952 Life-zone map. Syst. Zool., 1:24-25.

Mulch, E. E. and W. C. Gamble

1954 Game fishes of Arizona. Ariz. Game and Fish Dept.

Muller, C. H.

1939 Relations of the vegetation and climatic types in Nuevo Leon, Mexico. Amer. Midl. Nat., 21:687-729.

1940 Plant succession in the *Larrea-Flourensia* climax. Ecology, 21:206-212.

Munns, E. N.

1938 The distribution of important forest trees of the United States. U. S. Dept. Agr., Misc. Publ., 287:1-176.

Munz, P. A. and D. D. Keck

1949 California plant communities. El Aliso, 2:87-105, 199-202.

1959 A California flora. Univ. Calif. Press, Berkeley.

Murie, O. J.

1951 The elk of North America. Wildlife Mgmt. Inst., Washington, D. C.

Murray, A. V.

1959 An analysis of changes in Sonoran Desert vegetation for the years 1928-1957. Master's thesis, The University of Arizona, Tucson.

Nelson, E. W.

1922 Lower California and its natural resources. Mem. [U. S.] Nat. Acad. Sci., 16: 1-194.

Nichol, A. A.

1937 The natural vegetation of Arizona. Univ. Ariz. Agr. Exp. Sta. Tech. Bull., 68:181-222. 2nd ed., 1952, Tech. Bull., 127:189-230.

Nikiforoff, C. C.

1937 General trends of the desert type of soil formation. Soil Sci., 43: 105-131.

Norris, J. J.

1950 Effect of rodents, rabbits, and cattle on two vegetation types in semi-desert range land. New Mexico Agr. Exp. Sta. Bull. 353:1-24.

Norris, K. S.

1958 The evolution and systematics of the iguanid genus *Uma* and its relation to the evolution of other North American reptiles. Bull. Amer. Mus. Nat. Hist., 114:251-326.

Odum, E. P.

1945 The concept of the biome as applied to the distribution of North American birds. Wilson Bull., 57:191-201.

1959 Fundamentals of ecology. 2nd ed. W. B. Saunders Co., Philadelphia.

Olin, G.

1959 Mammals of the southwest deserts. 2nd ed. Southwestern Monuments Assoc., Pop. Ser., 8:1-112.

1961 Mammals of the southwest mountains and mesas. Southwestern Monuments Assoc., Pop. Ser., 9:1-126.

Osgood, W. H.

1900 Revision of the pocket mice of the genus Perognathus. U. S. Dept. Agr., N. Amer. Fauna, 18:1-73.

Packard, A. S. Jr.

1878 Some characteristics of the central zoo-geographical province of the United States. Amer. Nat., 12:512-517.

Parker, D.

1944 Review of Dice's "Biotic provinces of North America." Amer. Midl. Nat., 32:254.

Parker, K. W.

1945 Juniper comes to the grasslands: why it invades southwestern grassland — suggestions on control. Amer. Cattle Producer, 27:12-14, 30-32.

Parker, K. W. and S. C. Martin

1952 The mesquite problem on southern Arizona ranges. U. S. Dept. Agr. Circ. 908:1-70.

Patraw, P. M.

1931 Plant succession on Kaibab limestone. Grand Canyon Nat. Notes, 6:6-9.

1953 Flowers of the southwest mesas. Southwestern Monuments Assoc., Pop Ser., 5:1-112.

Pearson, G. A.

1920a Factors controlling the distribution of forest types, Part I. Ecology, 1:139-159.

1920b Factors controlling the distribution of forest types, Part II. Ecology, 1:289-308.

Pearson, G. A.

1930 Studies of climate and soil in relation to forest management in the Southwestern United States. Ecology, 18:139-144.

1931 Forest types in the Southwest as determined by climate and soil. U. S. Dept. Agr. Tech. Bull. 247:1-144.

1933 The conifers of northern Arizona. Plateau, 6:1-7.

1942 Herbaceous vegetation a factor in natural regeneration of ponderosa pine in the Southwest. Ecol. Monog., 12:315-338.

1950 Management of ponderosa pine in the Southwest. U. S. Dept. Agr., Forest Serv. Monog. 6:1-218.

Peattie, D. C.

1953 A natural history of western trees. Houghton-Mifflin, Boston.

Peters, J. A.

1955 Use and misuse of the biotic province concept. Amer. Nat., 89:21-28.

Peterson, R. T.

1941 A field guide to western birds. Houghton-Mifflin, Boston.

1942 Life-zones, biomes, or life-forms? Audubon Mag., 44:21-30.

Phillips, A. R.

1939 The faunal areas of Arizona, based on bird distribution. Master's thesis, The University of Arizona.

1956 The migrations of birds in Northern Arizona. Plateau, 29:31-35.

1959 The nature of avain species. *In* Species: Modern concepts. J. Ariz. Acad. Sci., 1:22-30.

Phillips, A. R. and G. Monson

1963 An annotated check list of the birds of Arizona. *In* The Vertebrates of Arizona. Univ. Ariz. Press, Tucson.

Pitelka, F. A.

1941 Distribution of birds in relation to major biotic communities. Amer. Midl. Nat., 25:113-137.

1943 Review of Dice's "Biotic Provinces of North America." Condor, 45:203-204.

Preble, E. A.

1923 A biological survey of the Pribilof Islands, Alaska. Life-zone relationships. U.S. Dept. Agr., N. Amer. Fauna, 46:1-255.

Preston, R. J., Jr.

1947 Rocky Mountain trees. 2nd ed., Ames, Iowa.

Rasmussen, D. I.

1941 Biotic communities of Kaibab Plateau, Arizona. Ecol. Monog., 11:229-275.

Raup, H. M.

1951 Vegetation and cryoplanation. Ohio J. Sci., 51:105-116.

Read, A. D.

1915 The flora of the Williams Division of the Tusayan National Forest, Arizona. Plant World, 18:112-123.

Reed, T. A.

1933 The North American high-level anticyclone. Monthly Weather Rev., 61: 321-325.

1939 Thermal aspects of the high-level anticyclone. Monthly Weather Rev., 67.

Reynolds, H. G. and J. W. Bohning

1956 Effects of burning on a desert grass-shrub range in southern Arizona. Ecology, 37:769-777.

Robbins, W. W.

1917 Native vegetation and climate of Colorado in their relation to agriculture. Colorado Agr. Exp. Sta. Bull. 224:1-56.

Robinson, H. H

1913 The San Franciscan volcanic field, Arizona. U. S. Geol. Surv. Prof. Paper 76:1-213.

Rusby, H. H.

1889 General floral features of the San Francisco and Mogollon Mountains of Arizona and New Mexico, and their adjacent regions. Trans. N. Y. Acad. Sci., 8:76-81.

Ruthven, A. G.

1907 A collection of reptiles and amphibians from southern New Mexico and Arizona. Bull. Amer. Mus. Nat. Hist., 23:483-604.

1908 The faunal affinities of the prairie region of central North America. Amer. Nat., 42:388-393.

Rydberg, P. A.

1914 Phytogeographical notes on the Rocky Mountain region. II. Origin of the alpine flora. Bull. Torrey Botan. Club, 41:89-103.

1916 Vegetative life-zones of the Rocky Mountain Region. Mem. N. Y. Bot. Garden, 6:477-499.

Rzedowski, J.

1956 Notas sobre la flora y la vegetación del estado de San Luis Potosí. III. Vegetación de la región de Guadalcazar. Anales del Instituto de Biología, México, D.F., 27:169-228.

Sampson, A. W.

1918 Climate and plant growth in certain vegetative associations. Bull. U. S. Dept. Agr., 700:1-72.

Sargent, C. S.

1926 Manual of the trees of North America (exclusive of Mexico). 2nd ed., Boston.

Saunders, A. A.

1921 A distributional list of the birds of Montana. Pac. Coast Avifauna, 14: 1-194.

Schroeder, W. L.

1953 History of juniper control on the Fort Apache Reservation. Ariz. Cattlelog, 8:18-25.

Schulman, E.

1956 Dendroclimatic changes in semiarid America. Univ. Ariz. Press, Tucson.

Schwalen, H. C.

1942 Rainfall and runoff in the upper Santa Cruz River drainage basin. Univ. Ariz. Agr. Exp. Sta. Tech. Bull. 95:421-472.

Sclater, P. L.

1858 On the general geographic distribution of the members of the class Aves. J. of Proc. Linnaean Soc. London (Zool.), 2:130-145.

Sellers, W. D.

1960a Arizona Climate. (Editor) Univ. Ariz. Press, Tucson.

1960b The climate of Arizona. *In* Arizona Climate, Univ. Ariz. Press, Tucson.

Shantz, H. L.

1936 Vegetation. *In* Arizona and its heritage. Univ. Ariz., Gen. Bull. 3:46-52.

Shantz, H. L. and R. L. Piemeisel

1925 Indicator significance of the natural vegetation of the southwestern desert region. J. Agr. Res., 28:721-801.

Shantz, H. L. and R. Zon

1924 Natural vegetation. U. S. Dept. Agr., Atlas of Amer. Agr., pt. 1, sec. E. Washington.

Shaw, C. H.

1909 The causes of timber line on mountains; the role of snow. Plant World, 12:169-181.

Shelford, V. E.

1932 Life-zones, modern ecology, and the failure of temperature summing. Wilson Bull., 44:144-157.

1945 The relative merits of the life zone and biome concepts. Wilson Bull., 57:248-252.

Shreve, F.

1911 The influence of low temperature on the distribution of the giant cactus. Plant World, 14:136-146.

1914 A montane rain-forest. A contribution to the physiological plant geography of Jamaica. Carnegie Inst. Washington, Publ. 199:1-110.

1915 The vegetation of a desert mountain range as conditioned by climatic factors. Carnegie Inst. Washington, Publ. 217:1-112.

1917a The physical control of vegetation in rain-forest and desert mountain. Plant World, 20:135-141.

1917b A map of the vegetation of the United States. Geog. Rev., 3:119-125.

1919 A comparison of the vegetational features of two desert mountain ranges. Plant World, 22:291-307.

1922 Conditions indirectly affecting vertical distribution on desert mountains. Ecology, 3:269-274.

1924 Soil temperature as influenced by altitude and slope exposure. Ecology, 5:128-136.

1925 Ecological aspects of the deserts of California. Ecology, 6:93-103.

1929 Changes in desert vegetation. Ecology, 10:364-373.

1934 Vegetation of the northwestern coast of Mexico. Bull. Torrey Botan. Club, 61:373-380.

1936 The plant life of the Sonora Desert. Sci. Monthly, 42:195-213.

1939 Observations on the vegetation of Chihuahua. Madrono, 5:1-13.

1942a The vegetation of Arizona. *In* Flowering plants and ferns of Arizona, by T. H. Kearney and R. H. Peebles. U. S. Dept. Agr. Misc. Publ., 423:10-23.

1942b The desert vegetation of North America. Botan. Rev., 8:195-246.

1942c Grassland and related vegetation in northern Mexico. Madroño, 6:190-198.

1944 Rainfall of northern Mexico. Ecology, 25:105-111.

1951 Vegetation and flora of the Sonoran Desert. Vol. I, Vegetation. Carnegie Inst. Washington, Publ. 591:1-192.

Shreve, F. and A. L. Hinckley

1937 Thirty years of change in desert vegetation. Ecology, 18:463-478.

Smith, A. P.

1908 Some data and records from the Whetstone Mountains, Arizona. Condor, 10:75-78.

Smith, H. M.

1940 An analysis of the biotic provinces of Mexico, as indicated by the distribution of the lizards of the genus *Sceloporus*. Anales Escuela Nac. Ciencias Biol., Mexico, 1:96-110.

Smith, H. V.

1956 The climate of Arizona. Univ. Ariz. Agr. Exp. Sta. Bull. 279:1-99.

Soper, J. D.

1946 Mammals of the northern Great Plains along the international boundary in Canada. J. Mammal., 27:127-153.

Spaulding, V. M.

1909 Distribution and movements of desert plants. Carnegie Inst. Washington, Publ. 113:1-144.

1910 Plant associations of the Desert Laboratory domain and adjacent valley. Plant World, 13:31-42.

Spangle, P. and M. Sutton

1949 The botany of Montezuma Well. Plateau, 22:11-19.

Standley, P. C.

1915 Vegetation of the Brazos Canyon, New Mexico. Plant World, 18:179-191.

Stebbins, R. C.

1949 Speciation in salamanders of the plethodontid genus *Ensatina*. Univ. Calif. Publ. Zool., 48:377-526.

1954 Amphibians and reptiles of Western North America. McGraw-Hill Book Co., New York.

Stejneger, L.

1893 Annotated list of the reptiles and batrachians collected by the Death Valley Expedition in 1891, with descriptions of new species. *In* U. S. Dept. Agr., N. Amer. Fauna, 7:159-228.

Stevens, J. S.

1905 Life areas of California. Trans. San Diego Soc. Nat. Hist., 1:1-25.

Storer, T. I. and R. L. Usinger

1957 General Zoology. McGraw-Hill Book Co., New York.

Sturdevant, G. E.

1927 Flora of the Tonto Platform. Grand Canyon Nat. Notes, 1:1-2.

Sudworth, G. B.

1915 The cypress and juniper trees of the Rocky Mountain region. U. S. Dept. Agr. Bull., 207:1-36.

1934 Poplars, principal tree willows and walnuts of the Rocky Mountain region. U. S. Dept. Agr. Bull., 680:1-45.

Sutton, M.

1952 A botanical reconnaissance in Oak Creek Canyon. Plateau, 25:30-42.

Sutton, G. M. and A. R. Phillips

1942 June bird life of the Papago Indian Reservation, Arizona. Condor, 44: 57-65.

Swarth, H. S.

1914 A distributional list of the birds of Arizona. Pac. Coast Avifauna, 10: 1-133.

1920 Birds of the Papago Saguaro National Monument and the neighboring region, Arizona. U. S. Dept. Interior, National Park Service. 63 p.

1929 Faunal areas of southern Arizona; a study in animal distribution. Proc. Calif. Acad. Sci., 18:267-383.

Sykes, G.

1931 Rainfall investigation in Arizona and Sonora by means of long-period rain gauges. Geog. Rev., 21:229-233.

Tansley, A. G.

1935 The use and abuse of vegetational concepts and terms. Ecology, 16: 284-307.

Taylor, W. P.

1922 A distributional and ecological study of Mt. Ranier, Washington. Ecology, 3:214-236.

Thornber, J. J.

1910 The grazing ranges of Arizona. Univ. Ariz. Agr. Exp. Sta. Bull., 65: 245-360.

1911 Plant acclimatization in southern Arizona. Plant World, 14:15-23.

Townsend, C. H. T.

1893 On the life zones of the Organ Mountains and adjacent region in southern New Mexico, with notes on the fauna of the range. Science, 22: 313-315.

1895-97 On the bio-geography of Mexico, Texas, New Mexico, and Arizona, with special reference to the limits of the life areas, and a provisional synopsis of the bio-geographic divisions of America. Trans. Texas Acad. Sci., 1 (1895) :71-96; 2 (1897) :33-86.

Turnage, W. V. and H. L. Hinkley

1938 Freezing weather in relation to plant distribution in the Sonoran Desert. Ecol. Monog., 8:529-550.

Turnage, W. V. and T. D. Mallory

1941 An analysis of rainfall in the Sonoran Desert and adjacent territory. Carnegie Inst. Washington, Publ. 529:1-110.

Turner, R. M.

1959 Evolution of the vegetation of the southwestern desert region. *In* University of Arizona, Arid Lands Colloquia, 1958-59:46-53.

U. S. Forest Service

1949 Areas characterized by major forest types in the United States (map). U. S. Dept. Agr. Washington, D. C.

van Rossem, A. J.

1931 Report on a collection of land birds from Sonora, Mexico. Trans. San Diego Soc. Nat. Hist., 6:237-304.

1932 The avifauna of Tiburon Island, Sonora, Mexico, with descriptions of four new races. Trans. San Diego Soc. Nat. Hist., 7:119-150.

1936a Notes on birds in relation to the faunal areas of south-central Arizona. Trans. San Diego Soc. Nat. Hist., 8:122-148.

1936b Birds of the Charleston Mountains, Nevada. Pac. Coast Avifauna, 24: 1-65.

1945 A distributional survey of the birds of Sonora, Mexico. Louisiana State Univ. Mus. Zool., Occ. Paper, 21:1-379.

Van Tyne, J. and A. J. Berger

1959 Fundamentals of ornithology. John Wiley and Sons, Inc., New York.

Vestal, A. G.

1914 Internal relations of terrestrial associations. Amer. Nat., 48:413-445.

Visher, S. S.

1924 Climatic laws. Ninety generalizations with numerous corollaries as to the geographic distribution of temperature, wind, moisture, etc. A summary of climate. John Wiley and Sons, Inc., New York.

Vorhies, C. T., R. Jenks and A. R. Phillips

1935 Bird records from the Tucson region, Arizona. Condor, 37:243-247.

Wallace, A. R.

1876 The geographical distribution of animals, with a study of the relations of living and extinct faunas as elucidating the past changes of the earth's surface. MacMillan and Co., London.

Wallmo, O. C.

1955 Vegetation of the Huachuca Mountains, Arizona. Amer. Midl. Nat., 54:466-480.

Weaver, J. E. and F. E. Clements

1938 Plant ecology. McGraw-Hill Book Co., New York.

Webb, W. L.

1950 Biogeographic regions of Texas and Oklahoma. Ecology, 31:426-433.

Webster, G. L.

1961 The altitudinal limits of vascular plants. Ecology, 42:587-590.

Whitfield, C. J. and E. L. Beutner

1938 Natural vegetation in the desert plains grassland. Ecology, 19:26-37.

Whitfield, C. J. and H. L. Anderson

1938 Secondary succession in the desert plains grassland. Ecology, 19:171-180.

Whittaker, R. H.

1951 A criticism of the plant association and climatic climax concepts. Northwest Sci., 25:17-31.

1953 A consideration of climax theory: The climax as a population and pattern. Ecol. Monog., 23:41-78.

1956 Vegetation of the Great Smoky Mountains. Ecol. Monog., 26:1-80.

1957 Recent evolution of ecological concepts in relation to the eastern forests of North America. Amer. J. Botan., 44:197-206.

1960 Vegetation of the Siskiyou Mountains, Oregon and California. Ecol. Monog., 30:275-338.

Willet, G.

1933 A revised list of the birds of southwestern California. Pac. Coast Avifauna, 21:1-204.

Woodbury, A. M.

1947 Distribution of pigmy conifers in Utah and northeastern Arizona. Ecology, 28:113-126.

1954 Principles of general ecology. The Blakiston Co., New York.

Woodbury, A. M. and H. N. Russell, Jr.

1945 Birds of the Navajo Country. Univ. Utah Bull. 35, Biol. Ser. 9(1): 1-160.

Woodin, H. E. and A. A. Lindsey

1954 Juniper-piñon east of continental divide, as analyzed by the line-strip method. Ecology, 35:473-489.

Woolsey, T. S.

1911 Western yellow pine in Arizona and New Mexico. U. S. Dept. Agr. Forest Serv. Bull., 101:1-33.

Wooton, E. O. and P. C. Standley

1913 Description of new plants, preliminary to a report on the flora of New Mexico. Contrib. U. S. Nat. Herbarium, 16:109-196.

Yang, T. W.

1957 Vegetational, edaphic, and faunal correlations of the western slope of the Tucson Mountains and the adjoining Avra Valley. Ph.D. Dissertation, The University of Arizona.

1961 The recent expansion of creosotebush (*Larrea divaricata*) in the North American desert. Western Reserve Academy, Nat. Hist. Mus., 1:-1-11.

Yang, T. W. and C. H. Lowe

1956 Correlation of major vegetation climaxes with soil characteristics in the Sonoran Desert. Science, 123:542.

York, C. L.

1949 The physical and vegetational basis for animal distribution in the Sierra Vieja Range of southwestern Texas. Tex. J. Sci., 3:46-62.

Zweifel, R. S.

1962 Analysis of hybridization between two subspecies of the desert whip-tailed lizard, *Cnemidophorus tigris*. Copeia, 1962 (4):749-766.

PART 2

FISHES OF ARIZONA

Robert Rush Miller
The University of Michigan
and
Charles H. Lowe
The University of Arizona

Although published references on the fishes of Arizona go back more than a century, to 1848, no listing of the state's fish fauna has previously appeared. The present check list is an inventory of the species of fishes, known to us, that occur in Arizona waters. It is based on extensive field work and review of the literature and includes both the native kinds as well as those introduced by various agencies and individuals for various reasons, primarily for some aspect of sport fishing.

During the century 1850-1960, striking changes took place in the aquatic habitats and in the fishes of Arizona, and this upset of natural conditions is continuing undiminished. Contraction and elimination of surface flows have restricted distributions, and the stocking of exotic species has resulted in replacement as well as reduction of native kinds (Miller, 1961). Although no indigenous fish is yet known to have been eliminated from the state, the following species are nearing extinction in Arizona: the native trout (*Salmo gilae*), humpback chub (*Gila cypha*), Colorado squawfish (*Ptychocheilus lucius*), Little Colorado spinedace (*Lepidomeda vittata*, restricted to Arizona), desert pupfish (*Cyprinodon macularius*), and Gila topminnow (*Poeciliopsis occidentalis*).

The number of exotic species known from Arizona has increased rapidly in the past 12 years. Just before 1950, there were more native than introduced species, 25 versus 19 (Miller, 1949). By 1960, however, the total estimated fauna of 61 species included 33 alien forms (Lowe, 1960). The total number of species given in the present list is 68, of which 27 are native, 37 are established exotics and 4 are hypothetical aliens which may become permanent residents before this paper is published. It is evident that the number of introductions will continue to grow. Those fishes to be included in the fauna in the near future will include species now in the process of becoming successfully established and others yet to be planted for one reason or another. It is unfortunate that man's modification of his environment has such deleterious effects on the native fishes, since this fauna comprises a group of animals of considerable interest and scientific importance that are yet inadequately studied.

The general plan for the check lists of vertebrates in this series is the same, and documentation is but rarely given within the lists. However, because this is the first such compilation for fishes it is appropriate to cite here some of the pertinent literature, especially since this is widely scattered.

The first important fish collections from Arizona were made by personnel of the U. S. and Mexican Boundary Survey during the middle of the nineteenth century (1851-54). The fishes were reported on by Baird and Girard and by Girard (1859, with references therein to earlier papers). Nearly 20 years followed before Cope and Yarrow (1875) wrote an important paper that dealt in part with fishes from Arizona. In their historical review and check list, Evermann and Rutter (1895) summarized what was then known of the fish fauna of the entire Colorado River basin. This was followed with a valuable contribution by Gilbert and Scofield (1898) covering species from the Gila and lower Colorado rivers. Snyder (1915) described collections made during the Mexican Boundary resurvey (1892-94), including significant data on habits and life colors by the collector, Dr. Edgar A. Mearns. More recent contributions that treat the Arizona fish fauna in varied detail are by the following, arranged chronologically: Hubbs and Miller (1941), Hubbs, Hubbs, and Johnson (1943), Miller (1943, 1945, 1946, 1952, 1955, 1959, 1963), Dill (1944), Wallis (1951), Hubbs and Miller (1953), Hubbs (1954), Winn and Miller (1954), Shapovalov, Dill, and Cordone (1959), and Miller and Hubbs (1960). We have also drawn on a number of other papers, several of which have appeared in *California Fish and Game* since Glidden's (1941) report of *Elops affinis* from the lower Colorado River. The name lower Colorado River in the present work refers to the section of the river below Lake Mead.

In the check list we employ the common names given in the second edition (1960) of "A List of Common and Scientific Names of Fishes from the United States and Canada," published by the American Fisheries Society as Special Publication 2, 102 pages. Exceptions are for *Gila robusta* and *Fundulus zebrinus;* reasons for using the vernaculars given here for these two species are cited by Sigler and Miller (1963).

It is a pleasure to acknowledge the considerable and long-time help of A. W. Yoder, former Chief of Fisheries of the Arizona Game and Fish Department, and, more recently, the assistance of David I. Foster of that department, and Dr. William J. McConnell of the University of Arizona (Cooperative Wildlife Research Unit and Department of Zoology). The senior author's field work in Arizona (1939-1962) has been supported by grants from the Horace H. Rackham School of Graduate Studies and the Museum of Zoology, the University of Michigan, and from the National Science Foundation (G-12904, G-15914).

LITERATURE CITED

Cope, E. D., and H. C. Yarrow

1875 Report upon the collections of fishes made in portions of Nevada, Utah, California, Colorado, New Mexico, and Arizona, during the years 1871, 1872, 1873, and 1874. Rept. Geog. Geol. Expl. Surv. W. 100th Mer. (Wheeler Surv.) 5:635-703.

Dill, W. A.

1944 The fishery of the lower Colorado River. Calif. Fish and Game, 30(3): 109-211.

Evermann, B. W., and C. Rutter

1895 The fishes of the Colorado Basin. Bull. U.S. Bur. Fish., 14 (1894): 473-486.

Gilbert, C. H., and N. B. Scofield

1898 Notes on a collection of fishes from the Colorado basin in Arizona. Proc. U.S. Nat. Mus., 20:487-499.

Girard, C.

1859 Ichthyology of the Boundary. U.S. and Mex. Boundary Surv., 2(2):1-85.

Glidden, E. H.

1941 Occurence of *Elops affinis* in the Colorado River. Calif. Fish and Game, 27(4):272-273.

Hubbs, C. L.

1954 Establishment of a forage fish, the red shiner *(Notropis lutrensis)*, in the lower Colorado River system. Calif. Fish and Game, 40(3): 287-294.

Hubbs, C. L., L. C. Hubbs, and R. E. Johnson

1943 Hybridization in nature between species of catostomid fishes. Contrib. Lab. Vert. Biol., Univ. Mich., 22:1-76.

Hubbs, C. L., and R. R. Miller

1941 Studies of the fishes of the Order Cyprinodontes. XVII. Genera and species of the Colorado River system. Occ. Papers Mus. Zool. Univ. Mich., 433:1-9.

1953 Hybridization in nature between the fish genera *Catostomus* and *Xyrauchen*. Papers Mich. Acad. Sci., Arts, and Lett., 38(1952): 207-233.

Lowe, C. H.

1960 Fishes. *In* Arizona, its people and resources. Univ. Ariz. Press, Tucson, 171-172.

Miller, R. R.

1943 The status of *Cyprinodon macularius* and *Cyprinodon nevadensis*, two desert fishes of western North America. Occ. Papers Mus. Zool. Univ. Mich., 473:1-25.

1945 A new cyprinid fish from southern Arizona, and Sonora, Mexico, with the description of a new subgenus of *Gila* and a review of related species. Copeia, 1945, 2:104-110.

1946 *Gila cypha*, a remarkable new species of cyprinid fish from the Colorado River in Grand Canyon, Arizona. J. Wash. Acad. Sci., 36(12):409-415.

1949 Keys for the identification of the fishes of Arizona. Univ. Mich. Mus. Zool., 1949:1-8 (mimeo).

1952 Bait fishes of the lower Colorado River from Lake Mead, Nevada, to Yuma, Arizona, with a key for their identification. Calif. Fish and Game, 38(1): 7-42.

Miller, R. R.

1955 Fish remains from archaeological sites in the lower Colorado River basin, Arizona. Papers Mich. Acad. Sci., Arts, and Lett., 40(1954):125-136.

1959 Origin and affinities of the freshwater fish fauna of western North America. *In*: Zoogeography, C. L. Hubbs, ed. Amer. Assoc. Adv. Sci., Publ. 51(1958):187-222.

1961 Man and the changing fish fauna of the American Southwest. Papers Mich. Acad. Sci., Arts, and Lett., 46(1960):365-404.

1963 Distribution, variation, and ecology of *Lepidomeda vittata*, a rare cyprinid fish endemic to eastern Arizona. Copeia, 1963, 1:1-5.

Miller, R. R., and C. L. Hubbs

1960 The spiny-rayed cyprinid fishes (Plagopterini) of the Colorado River system in western North America. Misc. Publ. Mus. Zool. Univ. Mich., 115:1-39.

Robins, C. R., and E. C. Raney

1957 Distributional and nomenclatorial notes on the suckers of the genus *Moxostoma*. Copeia, 1957, 2:154-155.

Shapovalov, L., W. A. Dill, and A. J. Cordone

1959 A revised check list of the freshwater and anadromous fishes of California. Calif. Fish and Game, 45(3):159-180.

Sigler, W. F. and R. R. Miller

1963 Fishes of Utah. Utah St. Dep. Fish and Game, 1963:1-204.

Snyder, J. O.

1915 Notes on a collection of fishes made by Dr. Edgar A. Mearns from rivers tributary to the Gulf of California. Proc. U.S. Nat. Mus., 49:573-586.

Wallis, O. L.

1951 The status of the fish fauna of the Lake Mead National Recreational Area, Arizona-Nevada. Trans. Amer. Fish. Soc., 80(1950):84-92.

Winn, H. E., and R. R. Miller

1954 Native postlarval fishes of the lower Colorado River basin, with a key to their identification. Calif. Fish and Game, 40(3):273-285.

Family Elopidae: Tarpons

1. ***Elops affinis*** Regan. Machete; Tenpounder. Native. This species has been taken sporadically in the lower Colorado River where Imperial Dam evidently represents the physical limit of its upstream occurrence. It was first recorded in the river in 1941, and very few captures have been made since then. The species ranges from Salton Sea, California, to northern Peru.

Family Clupeidae: Herrings

2. ***Dorosoma (Signalosa) petenense*** (Günther). Threadfin Shad. Introduced. Introduced as a forage fish into Lake Havasu, on the lower Colorado River, in 1954 and 1955. Brood stock came from the Tennessee River below Watts Bar Dam. It is now abundant in the Colorado and has also been successfully introduced into warm-water lakes in Arizona below 5,000 feet elevation.

Family Salmonidae: Salmons, Trouts, and Graylings

[***Oncorhynchus nerka*** (Walbaum). Sockeye Salmon. Introduced. Hypothetical. The kokanee (or little redfish), a dwarfed landlocked form of this salmon, was first introduced by the Arizona Game and Fish Department into Ashurst Lake in 1959 and is now present in Ashurst and Luna Lakes. The possibility of reproduction, when the fish become adult, is believed to be remote.]

3. ***Salmo gilae*** Miller. Gila Trout. Native. The native Arizona trout occurs in tributaries of Salt River, in the White Mountains, and in remote headwaters of Gila River (Eagle Creek, Greenlee County); it also persists in a few tributaries of this river in New Mexico. Until the early 1900's it occurred in Blue River (tributary of Gila River in Greenlee County) and in Oak Creek (south of Flagstaff). This fish was also introduced via a canal into the headwaters of the Little Colorado River. Although now extinct over most of its original range, this depleted and endangered trout may still be found at elevations above 10,000 feet in the Salt River drainage and in tributaries to Eagle Creek. The Arizona Game and Fish Department, alarmed over the rapid disappearance of the species, is attempting to restore this native fish through artificial propagation and restocking.

4. ***Salmo clarki*** Richardson. Cutthroat Trout. Introduced. As determined through extensive inquiry and examination of museum specimens, no native cutthroat trout existed within historic time in Arizona. Early records of Colorado River cutthroat (*S. clarki pleuriticus* Cope), from White River and Little Colorado River in Arizona, represent misidentifications for *S gilae.* Cutthroat trout, probably Yellowstone cutthroat, *S. clarki lewisi* (Girard), for the most part, have been stocked in suitable cold-water streams and lakes in Arizona for many years.

5. *Salmo gairdneri* Richardson. Rainbow Trout. Introduced. First introduced into Arizona around the turn of the century, this species has been raised in hatcheries and widely stocked since then in suitable waters from the lower Colorado River to the White Mountains. It hybridizes readily with the cutthroat trout and also with the Arizona native trout, to their detriment.

6. *Salmo trutta* Linnaeus. Brown Trout. Introduced. Planted from hatchery stock in suitable cold-water streams and lakes in mountain areas above about 6,000 feet elevation.

7. *Salvelinus fontinalis* (Mitchill). Brook Trout. Introduced. Occasionally reared from various hatchery stocks and sporadically planted in the more remote high mountain trout waters of the state.

8. *Thymallus arcticus* (Pallas). Arctic Grayling. Introduced. Planted in Big Lake, *ca.* 9,000 feet elevation, Apache County, in 1943 (A. W. Yoder, pers., 1950). The stock came from Grebe Lake in Yellowstone National Park and is assignable to *T. arcticus tricolor* Cope.

Family Cyprinidae: Minnows

9. *Cyprinus carpio* Linnaeus. Carp. Introduced. This Asiatic species, a stranger in the Colorado River at Yuma in 1890, is now abundant in warm waters throughout the state.

10. *Carassius auratus* (Linnaeus). Goldfish. Introduced. While frequently used as a baitfish along the lower Colorado River, no captures from the river have been reported. Goldfish were collected in Whitewater Creek near Douglas, Cochise County, in 1939 and 1943, and an adult was captured in Babocomari Creek, Cochise County, in 1950. In that year goldfish were reported in San Carlos Lake (reservoir on Gila River formed by Coolidge Dam) and in 1953 an adult was taken by hook and line at Canyon Lake on Salt River.

11. *Notemigonus crysoleucas* (Mitchill). Golden Shiner. Introduced. This species was well established in Lake Mary, southeast of Flagstaff, and in Mormon Lake, in 1934. It also has been used as a baitfish along the lower Colorado River.

12. *Gila atraria* (Girard). Utah Chub. Introduced. The listing of this species as part of the fish fauna of Arizona rests on its reported occurrence in the Colorado River below Hoover Dam. It is being used as a baitfish along the lower Colorado and is established at a number of places elsewhere in the Colorado River basin, e.g., in Aztec Creek north of Rainbow Bridge National Monument, San Juan County, Utah, where it was collected in 1958.

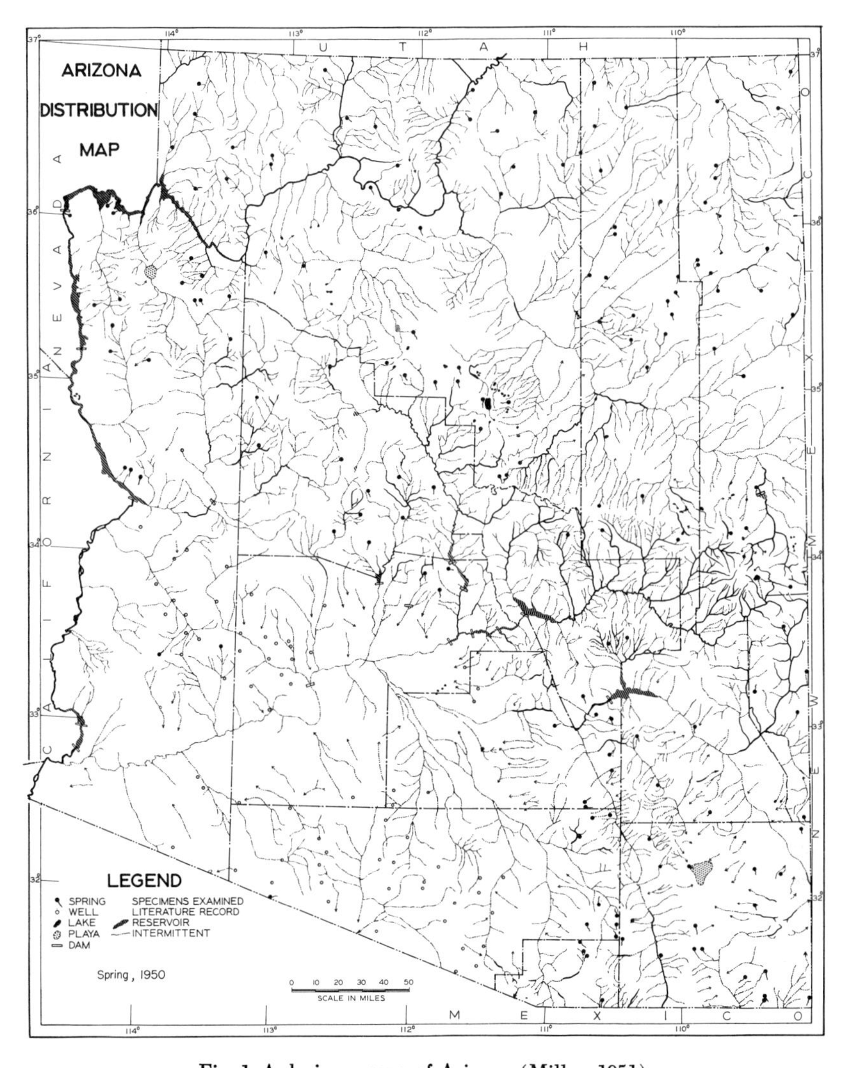

Fig. 1. A drainage map of Arizona (Miller, 1951).

13. *Gila robusta* Baird and Girard. Colorado Chub. Native. Widely distributed in rivers and streams of the state in warm waters, where it is represented by three well-defined subspecies, as follows: *G. r. intermedia* (Girard), the smallest of the three, is restricted to the Gila River basin where it lives in the smaller streams and in springs. *G. r. robusta* Baird and Girard, the roundtail, generally inhabits major tributaries of the Gila and Colorado rivers and attains a length of about 15 inches; it is often called Verde trout by residents. The bonytail, *G. r. elegans* Baird and Girard, is now evidently extinct in the Gila River drainage (last capture known to us was made in 1926 about 10 miles above the mouth of the Gila) and is becoming scarce in the Colorado River. In 1904, it was the most abundant fish at Yuma. This extreme form of *Gila robusta* is especially adapted by its streamlined body, large falcate fins, and reduced scales, for life in swift waters; hence it occurs almost exclusively in the main channel of the Colorado River. It may reach a length of nearly 17 inches and weigh between one and two pounds.

14. *Gila cypha* Miller. Humpback Chub. Native. This bizarre species, with its prominent nuchal hump and almost scaleless, streamlined body, was described in 1946 from near the mouth of Bright Angel Creek on the floor of Grand Canyon. No further specimens have been captured in Arizona although the species once occurred in Boulder Canyon and still may be caught in the upper Colorado River basin in Utah. Its relationship to *Gila robusta* is currently under study.

15. *Gila purpurea* (Girard). Yaqui Chub. Native. This species barely enters the Arizona fauna at San Bernardino Ranch, 18-19 miles east of Douglas, in the Yaqui River basin, where it was first collected in 1854. In 1896 it was recorded also from Morse Canyon, an upper tributary of Turkey Creek on the east side of the enclosed Cochise basin north of Douglas. Unfortunately the specimens have been lost and the identification cannot be verified; the species no longer inhabits that stream. The Yaqui chub, as the name implies, occurs in the Yaqui drainage, and also in the Río Sonora, in Sonora, Mexico.

16. *Gila ditaenia* Miller. Sonora Chub. Native. Known in Arizona only from Sycamore Canyon Creek, 14 miles west of Nogales, where it is the only species of native fish present in the small, permanent spring-fed stream. It is restricted to the independent basin of Río de la Concepcíon, which lies mostly in northwestern Sonora, Mexico, and barely enters Arizona.

17. *Ptychocheilus lucius* Girard. Colorado Squawfish. Native. This large minnow, which once reached a weight of nearly 100 pounds and a length approaching 6 feet, has nearly disappeared from the Arizona fauna, although a few adults have been caught in recent years in the lower Colorado River and two young were collected in Salt River, near the mouth of

Cibecue Creek, Gila County, in 1950. A hundred years ago it was still common in San Pedro River (tributary to Gila River) near Charleston in Cochise County, where the residents called it "Salmon." The species is known only from the Colorado River basin, inhabiting the main channel and the larger tributaries.

18. *Rhinichthys osculus* (Girard). Speckled Dace. Native. Widespread in suitable habitats (usually above 3,500 feet elevation) in Arizona, and occurring over much of western North America. The species was described from Babocomari Creek, flood tributary of San Pedro River, just north of Fort Huachuca, where 10 specimens were obtained in 1851. Although the same stream was fished in the spring of 1904, the species was not obtained. In April, 1950, this intermittent stream was thoroughly explored but the only minnows taken were *Agosia chrysogaster* and *Gila robusta intermedia.* However, in other (permanent) tributaries of San Pedro River (Redfield Creek, Aravaipa Creek) several samples of the speckled dace have been taken at various times since 1950.

[***Gila (Snyderichthys) copei*** (Jordan and Gilbert). Leatherside Chub. Introduced. Hypothetical. This species has been used for bait on the lower Colorado River but apparently has not become established there although it now inhabits parts of the upper Colorado basin in Utah.]

19. *Agosia chrysogaster* Girard. Longfin Dace. Native. A common and usually tolerant native species inhabiting permanent streams in the desert and desert grassland of Arizona and New Mexico below 4,500 feet. Whether there is but one species of this genus in Arizona (Gila River basin, and Williams River drainage) or more than one has not been firmly established. An isolated population occurs in Cochise Basin and the genus is also known from the headwaters of the Yaqui River in southeastern Arizona. It is used as a baitfish.

20. *Tiaroga cobitis* Girard. Loach Minnow. Native. This is one of two endemic genera of the Gila River basin (*Meda* is the other) of Arizona, New Mexico, and northern Sonora, Mexico. Its distribution is becoming increasingly restricted because of elimination of suitable habitat over much of its range. This colorful little fish is found only on and near rocky riffles on which green algae are common.

21. *Notropis mearnsi* Snyder. Yaqui Shiner. Native. This shiner, the only native member in the western United States of an abundant eastern genus, occurs in Arizona only at San Bernardino Ranch, 18-19 miles east of Douglas, in the Yaqui River basin. Its specific separation within the *Notropis lutrensis* complex is in need of careful evaluation. As the common name suggests, this fish is widely distributed in the Yaqui River basin and also occurs in the Río Sonora, Sonora, Mexico.

22. *Notropis lutrensis* (Baird and Girard). Red Shiner. Introduced. This eastern bait and forage fish, the Plains red shiner (*N. l. lutrensis*), was established in the lower Colorado River in 1953. It is now widely distributed in warm-water streams, lakes, and ponds in Arizona.

23. *Notropis stramineus* (Cope). Sand Shiner. Introduced. This species was long known as *Notropis deliciosus* (Girard). It was collected by John T. Young of Winslow, on October 29, 1955, in Jacks Canyon, 4 miles east of Winslow, Navajo County. It was first observed by him at the mouth of nearby Chevelon Creek in 1938. The occurrence of this eastern shiner in Arizona is logically explained as an accidental planting when two shipments of largemouth bass from Dexter, New Mexico, were stocked in 1935 in Clear Creek and possibly also in Chevelon Creek. The unusual factual data on this introduction results from Mr. Young's natural curiosity, for which we are most grateful.

24. *Pimephales promelas* Rafinesque. Fathead Minnow. Introduced. The fathead minnow was first collected in Arizona on May 24, 1952, by William P. and Dorothy Knoch in Paría River at Lee's Ferry, Coconino County. Additional specimens were taken by Jack Hemphill (then of the Arizona Game and Fish Department) on October 2, 1952, from Big Lake (elev. *ca.* 9,000 feet) in the headwaters of the East Fork of Black River, and in Grand Canyon (Colorado River and Bright Angel Creek) and the Little Colorado River during August and September, 1960, by R. R. Miller and party. This species has also been used as a baitfish on the lower Colorado River.

25. *Campostoma ornatum* Girard. Mexican Stoneroller. Native. As in the case of *Gila purpurea* and *Notropis mearnsi,* this cyprinid occurs in Arizona only in the Yaqui basin (Rucker Canyon, Chiricahua Mountains; Whitewater Creek, near Douglas). It is widely distributed on both slopes of the Sierra Madre Occidental in Sonora, Chihuahua, and Durango, Mexico, and occurs also in the Big Bend region of Texas. *Campostoma pricei* Jordan and Thoburn, described from Rucker Canyon, is a synonym of *C. ornatum.*

26. *Lepidomeda mollispinis* Miller and Hubbs. Middle Colorado Spinedace. Native. The population occurring in Arizona is *L. mollispinis mollispinis* Miller and Hubbs, the Virgin River spinedace. As indicated by the common name, this fish is known only from the Virgin River drainage of Arizona, Nevada, and Utah. This and the following three species are members of a distinctive tribe, the Plagopterini, unique among New World minnows in possessing ossified, spine-like dorsal and pelvic rays.

27. *Lepidomeda. vittata* Cope. Little Colorado Spinedace. Native. We regret to have to report that this species has been wiped out over most of its original range (in the upper part of the Little Colorado River basin) and is now threatened with extinction. A thorough examination of the

Little Colorado River and most of its tributaries by C. L. Hubbs, R. R. Miller, and party in August 1960, failed to yield specimens. Only in one tributary, Clear Creek, is the species still known to persist (Miller, 1963).

28. *Meda fulgida* Girard. Spikedace. Native. This endemic species occurs only in the Gila River basin of Arizona and New Mexico. In the nineteenth century, when streams of the Gila basin had a more ample flow, the spikedace probably occurred also in Mexico, in the headwaters of the San Pedro River. The range has become highly restricted, but at a few localities the species remains locally abundant and easily seined, as does *Tiaroga cobitis; e.g.*, in now isolated and spring-fed Aravaipa Creek in the Galiuro Mountains, Pinal County, a permanent stream which today only reaches the now dry bed of San Pedro River at times of torrential flood. In the San Pedro it has also been collected in recent years (1954, 1956, 1959) by both of us, but only on a still-flowing short section of this barely surviving river near Charleston, Cochise County.

29. *Plagopterus argentissimus* Cope. Woundfin. Native. This, the most specialized species of the Plagopterini, originally lived in the lower Gila River but it has not been taken there since 1894 and it is extinct in that drainage. The woundfin still inhabits swift waters of the Virgin River basin in Arizona, Nevada, and Utah. It has been used as a baitfish on the lower Colorado River.

Family Catostomidae: Suckers[1]

30. *Ictiobus cyprinellus* (Valenciennes). Bigmouth Buffalo. Introduced. Common in Roosevelt and Apache Lakes in Salt River. Buffalofish (*Ictiobus sp.*) eggs are recorded as having been planted in Roosevelt Lake in 1918 by the U. S. Bureau of Fisheries. According to William Hulett, commercial fisherman interviewed at San Carlos Lake in 1950, this species attains a weight of 33 pounds and a length of about 30 to 35 inches depending on sex (males larger). The maximum weight of 33 pounds was again cited to us by T. T. Frazier, owner of the store and boat landing on Roosevelt Lake. Mr. Frazier seined 2,000 tons of this fish from Roosevelt Lake between October 1938 and May 1940.

31. *Carpiodes cyprinus* (LeSueur). Quillback. Introduced. Apache Lake, and probably Roosevelt Lake, on Salt River. Identified from a fresh adult, which was one of several gill-netted with bigmouth buffalos (*Ictiobus cyprinellus*) in Apache Lake in April 1952, by C. W. Mickel, commercial fisherman. A large specimen of each of the two species was brought to the University of Arizona by Robert Jantzen. Man-made Apache Lake

---

[1] The early record (1874) of an eastern sucker "*Moxostoma congestum*" from Ash Creek is discussed by Robins and Raney (1957) who concluded that this distributional record "must be considered in error since no species of *Moxostoma* is known from the area west of the Continental Divide and north of Mexico."

is one of a series of Salt River impoundments located downstream from Roosevelt Lake. It is possible that more than a single species of fish was represented in the sucker eggs planted in Roosevelt Lake in 1918 by the U. S. Bureau of Fisheries (see account of *Ictiobus cyprinellus*, above).

32. *Catostomus insignis* Baird and Girard. Sonora Sucker. Native. This is the common coarse-scaled sucker of the Gila River basin and it also inhabits the Williams River drainage, an eastern tributary of Colorado River in Mohave and Yavapai counties. The first record of the species from the Williams River drainage is based on specimens taken in 1947 by R. R. Miller and party from Burro Creek, about 20 miles southeast of Wikieup.

33. *Catostomus bernardini* Girard. Yaqui Sucker. Native. This coarse-scaled sucker, described in 1856, is known in Arizona only from San Bernardino Creek, 18-19 miles east of Douglas. This stream, a tributary of Yaqui River, is the type locality (just south of the International Boundary in Sonora, Mexico) for the species. The Yaqui sucker ranges southward in Pacific coastal streams of Sonora, Chihuahua, and Sinaloa, Mexico. It should be noted that two suckers collected in San Bernardino Creek by James R. Simon in 1943 and a series of large *Catostomus* taken there by C. H. Lowe and party on February 19, 1954, appear to be indistinguishable from *C. insignis*, the Sonora sucker.

34. *Catostomus latipinnis* Baird and Girard. Flannelmouth Sucker. Native. Like other big-channel fishes (*Gila robusta elegans, Ptychocheilus lucius*), this species is now scarce in Arizona. It became extinct at the type locality, the San Pedro River near Benson, before the turn of the century, and has been taken in recent years in the Gila basin only in Salt River above Roosevelt Lake. The species is common in the Colorado River and its larger tributaries above Hoover Dam. The flannelmouth sucker is used as a baitfish along the lower Colorado River and has recently become established in Lake Mohave where, on January 24, 1954, Al Jonez (of the Nevada Fish and Game Department) took one adult.

35. *Pantosteus clarki* (Baird and Girard). Gila Sucker. Native. This species, known only from the Gila River basin of Arizona and New Mexico, frequently hybridizes with *Catostomus insignis;* the two species often occur together in large numbers and are commonly taken in the same seine hauls.

36. *Pantosteus delphinus* (Cope). Bluehead Sucker. Native. This species occurs in Arizona in the Colorado River in Grand Canyon and in tributaries of the Colorado in northern and western Arizona. It has not been taken lower down in the drainage than the Williams River system (Burro Creek) of western Arizona. This is one of the native fishes that has been utilized as a baitfish along the Colorado. Its taxonomic status is currently under investigation.

37. *Xyrauchen texanus* (Abbott). Humpback Sucker. Native. The humpback was at one time common throughout the Colorado River basin, and was even marketed at Tombstone as "buffalo" around 1880. The species is still holding out in reservoirs of Salt River near Phoenix, and in the lower Colorado River. It hybridizes with *Catostomus insignis* and *Catostomus latipinnis.*

Family Ictaluridae: North American Catfishes

38. *Ictalurus punctatus* (Rafinesque). Eastern Channel Catfish. Introduced. Until recently this species was known as *Ictalurus lacustris* (Walbaum). The time and place of its introduction into Arizona is not definitely known, in part because the U. S. Bureau of Fisheries did not distinguish in their records between bullheads and channel catfish — all were merely called catfish. In 1950 we were told by R. C. Richardson, U. S. Bureau of Reclamation employee and long-time fisherman along the lower Colorado, that the first channel catfish were caught around Yuma about 1927. This statement receives independent support from Ralph B. Keeler (interviewed in 1950 at the Veterans Hospital in Whipple), who had lived on the river since 1929; he testified that both bullheads and channel catfish were present in the lower Colorado in 1929. He also stated, however, that catfish (kinds not specified) were reported to have been introduced around 1913, and others testify that the channel catfish was noted in the river in 1906. According to T. T. Frazier, Roosevelt Lake, *Ictalurus punctatus* was stocked in Salt River at the crossing of U. S. Highway 60 in 1938, and soon became abundant; he was positive that there were no channel catfish in Roosevelt Lake or in Salt River above the lake prior to 1938.

39. *Ictalurus pricei* (Rutter). Pacific Channel Catfish. Introduced (but perhaps once native). This species, the only freshwater catfish native to the Pacific slope of the Southwest, was planted on July 6, 1899, in a reservoir fed by Monkey Spring, about 7 miles NNW of Patagonia, Santa Cruz County. The stock presumably came from the Río Sonora basin of Sonora, Mexico, where the species still lives. It also may occur or may have occurred (in high water) as far up San Bernardino Creek as the ranch of the same name east of Douglas, since the type locality is just south of the International line; if so, it would also be native to Arizona (Yaqui River basin).

40. *Ictalurus melas* (Rafinesque). Black Bullhead. Introduced. This species is successful in many kinds of warm waters, from those as permanent as the Colorado River to those as transient as earthen dammed stock ponds that are recharged only by rainfall and run-off; the latter introductions are often on private lands. It was present in the Colorado River at Yuma in 1904, before channel catfish had become successfully established.

41. *Ictalurus nebulosus* (LeSueur). Brown Bullhead. Introduced. This species was first recorded in the lower Colorado River in the vicinity of Topock in 1910, but the specimen was subsequently lost and the identification remains unverified. Recently, however, this catfish was collected by J. B. Kimsey in this river; it is not known to occur elsewhere in Arizona, and is greatly outnumbered in the Colorado River by its close relative the black bullhead.

42. *Ictalurus natalis* (LeSueur). Yellow Bullhead. Introduced. This is the commonest bullhead in the lower Colorado River. Its origin in the Colorado is obscure, but its appearance there is thought to antedate that of the channel catfish.

43. *Pylodictis olivaris* (Rafinesque). Flathead Catfish. Introduced. Specimens of this species from Arizona were collected by R. R. Miller and party on May 2, 1950, in Bonita Creek, near its mouth in Gila River, 15 miles by road NNE of Safford, Graham County. This catfish has also been recorded from the Gila drainage of New Mexico. These represent the only specimen records we know for western North America. It has been recently planted (March, 1962) in six areas of the Colorado River north of Yuma by the Arizona Game and Fish Department. Flathead catfish are native from the upper Mississippi River basin and the Lake Michigan and Lake Erie drainages southward to northeastern Mexico. How and when the species gained access to Arizona is not known. W. J. McConnell (of the University of Arizona) recently informed us that this species has been reported by fishermen from San Carlos Lake (Coolidge Dam Reservoir) on Gila River, northwest and downstream from Safford. Thus the species may have been first introduced into San Carlos Lake, whence it made its way upstream to Bonita Creek.

Family Cyprinodontidae: Killifishes

44. *Fundulus zebrinus* Jordan and Gilbert. Southwestern Plains Killifish. Introduced. This fish inhabits the middle part of the Little Colorado River basin, where it is locally abundant. It was first observed at the mouth of Chevelon Creek; near Winslow, in 1938 by John T. Young, who collected a sample in 1955. The way in which the species arrived in Arizona is detailed under the account of *Notropis stramineus*. This killifish, native to parts of New Mexico and Texas, has been used for bait along the lower Colorado River.

45. *Cyprinodon macularius* Baird and Girard. Desert Pupfish. Native. This interesting killifish, once common in the lower Gila River basin, is now known in Arizona from only two localities: Monkey Spring, near Patagonia, and Quitovaquito, on the southern edge of Organ Pipe Cactus National Monument. The original description of this small species is based on eight specimens collected in 1851 in San Pedro River just below (north) the present town of Benson, Cochise County. Until 1950, however, no additional specimens had been secured from this stream. On April 21-22, 1950, the desert pupfish was seined by R. R. Miller, F. H. Miller, and H. E. Winn from Rio San Pedro, above upper Elias Dam, about 9 miles upstream (south) from San Pedro Ranch (about 8 miles south of the international boundary, elev. *ca.* 4,500 feet), Sonora, Mexico. This is about 50 airline miles south of Benson. The species occurs outside of Arizona in the Río Sonoyta in extreme northern Sonora, and in Baja California, Mexico, and in the Salton Sea basin, California.

Family Poeciliidae: Topminnows

46. *Gambusia affinis* (Baird and Girard). Mosquitofish. Introduced. The subspecies occurring here, *G. affinis affinis* (western mosquitofish), probably reached Arizona shortly after 1922, the year in which it was introduced into California. It was collected in 1926 from the Gila River near its mouth, and from Salt River between Phoenix and Tempe. It was seen in Lake Mead about 1938. It has thrived and has driven its native ecological counterpart, the next species listed, almost out of existence. It is abundantly successful in the highly turbid waters of dirt stock ponds as well as in clear water. The species is not native west of the continental divide.

47. *Poeciliopsis occidentalis* (Baird and Girard). Gila Topminnow. Native. This native livebearer, under pressure from introduced predators and competitors and the elimination of much of its habitat, is rapidly nearing extinction. Once widespread in the southern part of the Gila River basin, it is now known from only a few scattered localities. It was evidently completely replaced by *Gambusia a. affinis* in Arivaca Creek at Arivaca, southeastern Pima County, between April 12, 1957, and March 30, 1959. This was an unnecessary extermination since this species, like the better-known mosquitofish, is also effective in controlling mosquito larvae.

48. *Poeciliopsis sonoriensis* (Girard). Sonora Topminnow. Native. This fish now occurs in Arizona only at San Bernardino Ranch, 18-19 miles east of Douglas, in the Yaqui River basin. Its taxonomic status (as a species) is currently under investigation.

Family Serranidae: Sea Basses

49. *Roccus mississippiensis* (Jordan and Eigenmann). Yellow Bass. Introduced. Occurs in reservoir lakes in Salt River and Gila River. Locally known as striped bass and formerly called *Morone interrupta* Gill, this game fish was stocked in Roosevelt Lake on Salt River between 1929 and 1932, according to T. T. Frazier, owner and operator of the store and boat dock on the lake. Specimens were collected there in 1934 by a U. S. Bureau of Fisheries survey party. A stunted population soon developed in the lake and the species disappeared from this reservoir after a heavy flood about 1934. Mr. Frazier attributed the mass mortality to heavy silting. In 1940, Roosevelt Lake was emptied and since then (at least to 1953, when Frazier was last interviewed by us) no yellow bass have been seen in this reservoir. The species was very abundant in San Carlos Lake in the spring of 1950, but the fish only reached a weight of about one pound (as in Canyon Lake in 1953); William Hulett, local commercial fisherman, told us then that much larger individuals were taken in Sahuaro Lake. This species is native from the upper Mississippi basin south to the Tennessee River system in Alabama, and to Louisiana and eastern Texas.

50. *Roccus saxatilis* (Walbaum). Striped Bass. Introduced. Occurs in the lower Colorado River, where an experimental plant of 938 small individuals was made by the California Department of Fish and Game on April 15, 1959, immediately below the U. S. Highway 60 bridge near Blythe, California.

[***Roccus chrysops*** (Rafinesque). White Bass. Introduced. Hypothetical. In 1960 the Arizona Game and Fish Department first planted this species in Carl Pleasant Lake, in Agua Fria River.]

Family Centrarchidae: Sunfishes

51. *Micropterus dolomieui* Lacépède. Smallmouth Bass. Introduced. This bass was planted in southern Arizona about 1942, and is less widely distributed than the better-known largemouth bass. In 1956 a series was caught (hook and line) in Beaver Creek, Verde River drainage, and preserved for us by Robert Moses and party of the University of Arizona; they reported it to be abundant. It was first recorded from the lower Colorado River in 1953, and reported to have been stocked there at 4 miles below Parker Dam, in August, 1950.

52. *Micropterus punctulatus* (Rafinesque). Spotted Bass. Introduced. According to mimeographed reports of the Arizona Game and Fish Commission, spotted bass had evidently been introduced into the Verde River drainage as early as 1942. In 1949, large numbers were noted in West Clear Creek, a tributary of the Verde.

53. *Micropterus salmoides* (Lacépède). Largemouth Bass. Introduced. This well-known game fish is found in virtually all suitable warm-water lakes and reservoirs in the state including those of the Colorado, Gila, Salt, and Verde rivers; it occurs in the channels of these and other rivers, in numerous streams, and in large and small ranch ponds.

54. *Ambloplites rupestris* (Rafinesque). Rock Bass. Introduced. A small population of rock bass occurs in lower Oak Creek, in the Verde River drainage.

55. *Chaenobryttus gulosus* (Cuvier). Warmouth Bass. Introduced. In 1958 the warmouth was reported from the lower Colorado River. In August 1955, John H. Gerdes of the University of Arizona took an adult female by hook and line at Canyon Lake, which is one of the reservoirs below Roosevelt Lake on Salt River.

56. *Lepomis macrochirus* Rafinesque. Bluegill. Introduced. This is one of the most widely stocked panfish and is found in the same waters as catfishes, basses, crappies, and other popular warm-water species. It was established in Mormon Lake near Flagstaff in 1934, and in Salt River, near the upper end of Roosevelt Lake, in 1937. Bluegills are abundant in some of the remotest ranch stock ponds where they often remain as a crowded and stunted population following depletion of the largemouth bass that were originally planted in many such ponds as the popular bass-bluegill game and forage fish combination. The bluegill is present and often locally abundant in the reservoir lakes of Salt River, the lower Colorado River (and its irrigation canals, wherever water is checked) and its lakes, including Lake Mead; it also occurs in San Carlos Lake behind Coolidge Dam on Gila River, in the 20-mile stretch of the Gila below the dam, and in the main stream and lakes of the Verde River drainage including Oak Creek. In addition it is found in numerous small permanent streams throughout the state, in Peña Blanca Lake, Ruby Lake, Lake Carl Pleasant, White Horse Lake, upper and lower Lake Mary, Long Lake, Stoneman Lake, Stehr Lake, and many others. The first date of introduction of the bluegill to the fishery of the lower Colorado River is not known. The order of introduction of the species of panfish into the river, however, seems to have been first the green sunfish followed by the bluegill and, more recently, the redear sunfish followed by the warmouth.

57. *Lepomis microlophus* (Günther). Redear Sunfish. Introduced. The redear is established in the Colorado River and Gila River basins, including the Salt and Verde River drainages. It was stocked in Apache, Canyon, and Stewart Mountain Lakes as early as 1947, probably from the Texas State Hatchery at San Angelo where a brood stock of 70 fish was obtained late in 1946 (A. W. Yoder, pers., 1950). It was first caught in the lower Colorado River on April 27, 1951, where the species is now abundant.

[Hybrid sunfish, ***Lepomis machrochirus* × *Lepomis microlophus*.** Introduced. Hypothetical. Hybrids were stocked in 1947 in Encanto Lagoon, Canyon Lake, Hudspath Lake, Ashurst Lake and Willow Lake; and in 1948 in Carl Pleasant Diversion Dam. We are uncertain as to the current status of this fish.]

58. *Lepomis cyanellus* Rafinesque. Green Sunfish. Introduced. The green sunfish occurs in most of the lakes and streams where the bluegill is found. It is established in both the Colorado River and Gila River basins, including the Salt and Verde River drainages. While its advent in the lower Colorado River is unknown, according to residents it has been there much longer than the bluegill which, in recent years, has greatly outnumbered the green sunfish at most points where they occur together below

Lake Havasu. This species was collected in Salt River, between Phoenix and Tempe, and in Gila River near its mouth, in 1926, and seen in Lake Mead in 1942. The green sunfish's markedly strong preference for rocks and rocky situations may account for its lesser abundance than the bluegill in many lakes such as Apache, Canyon, Mary, Havasu, etc.

59. *Pomoxis annularis* Rafinesque. White Crappie. Introduced. In Arizona, as in eastern North America where crappie are native, this species is less common than the black crappie (next listed) and less widely introduced. Both crappies are panfish that have been planted in warm waters where bass, bluegill, and channel catfish are successful. The first report of the genus in the lower Colorado River was in 1934, when crappies (species not specified) were caught in Haughtelin Lake. The white crappie was reported in 1934 from Roosevelt Lake on Salt River, but we have not verified the identification to species.

60. *Pomoxis nigromaculatus* (LeSueur). Black Crappie. Introduced. This is the panfish most commonly taken in Arizona. It is prevalent, often to the point of great abundance, in all waters of the state in which other warm-water species such as bass, bluegill, and channel catfish are common. According to testimony from Ralph B. Keeler (Veterans Hospital, Whipple, 1950), he first noticed crappies (like this species) at Imperial Dam in 1941, but they were probably in the river a few years previously. Crappies were present in Mormon Lake near Flagstaff in 1934, and in San Carlos Lake in the Gila River in 1936. The black crappie is usually more plentiful even than the bluegill, and suffers similar stunting from overpopulation (crowding followed by starving) in many lakes and ponds.

Family Percidae: Perches

61. *Perca flavescens* (Mitchill). Yellow Perch. Introduced. In its native eastern range, from Kansas to South Carolina and northward to Canada, the yellow perch is a common inhabitant of lakes. It has been planted in a few of Arizona's lakes including most of those of Anderson Mesa (southeast of Flagstaff), where Lake Kinnikinick has had the largest population. It was present in Mormon Lake in 1934. The species also has been planted in Peck's Lake at Cottonwood, and in Lakeside Lake.

Family Mugilidae: Mullets

62. *Mugil cephalus* Linnaeus. Striped Mullet. Native. In Arizona this world-wide tropical and subtropical species is known only from the lower Colorado River which it has ascended at least to Imperial Dam, from the Gulf of California, where it is common. Dill (1944:163) wrote that "Some old residents report that it was always of sporadic occurrence in the river near Yuma, and did not become firmly established there until the extensive system of canals and drains was built." Its advent in the lower river may also be correlated with the decimation of the huge predatory Colorado squawfish (*Ptychocheilus lucius*), and the clearing up of the river after it was dammed.

Family Cottidae: Sculpins

[***Cottus bairdi*** Girard. Mottled Sculpin. Introduced. Hypothetical. The subspecies *Cottus bairdi semiscaber* (Cope), the Bonneville mottled sculpin, has been used for bait along the lower Colorado. It may have become established in the river as indicated by one or two sight records. However, no specimens are available to confirm these reports. The species is native to the upper Colorado River basin, well above Grand Canyon, and is common in the adjacent Bonneville basin of Utah and in the upper part of Snake River. It is widespread in eastern United States.]

Family Eleotridae: Sleepers

63. *Eleotris picta* Kner and Steindachner. Spotted Sleeper. Native. This essentially tropical species was caught by hook and line on April 16, 1952, in the canal spillway between Winterhaven and the Colorado River, California, just north of Yuma. It obviously reached this point by moving up the Colorado River from its mouth in the Gulf of California. No other records north of the Yaqui River, Sonora, Mexico, are known to us. The species ranges southward on the Pacific slope to Ecuador.

Addendum

64. *Mollienesia latipinna* (LeSueur). Sailfin Molly. Introduced. Abundant in the canals and general drainage system of Salt River in the vicinity of Phoenix, Maricopa County, and in Salt River upstream at least to Roosevelt Dam as well as in the reservoir.

PART 3

AMPHIBIANS AND REPTILES OF ARIZONA

Charles H. Lowe
The University of Arizona

This list is an inventory, with annotations, of the Recent amphibian and reptilian species known to occur in Arizona. The herpetofauna of the state is at this time and by this accounting comprised of 116 species: 22 of amphibians and 94 of reptiles. Two of these, the bullfrog and the soft-shell turtle, are not native; both are well established introductions to the fauna. It is well known that young alligators (*Alligator mississippiensis*), usually less than four feet in length, are often successfully introduced (from southeastern United States) for a period of a few years into Arizona ranch ponds which contain fish, frogs, etc. The alligator, however, has not become established anywhere in the state and is not included in this list.

Of the 22 species of amphibians, one is a native salamander which has a wide transcontinental distribution across North America (tiger salamander, *Ambystoma tigrinum*); the remainder are toads and frogs. Of the 94 species of reptiles, 5 are turtles, 41 are lizards, and 48 are snakes.

While both the amphibians and the reptiles are more prevalent in the deserts and grasslands throughout most of the state, several of them are primarily woodland species and some occur in coniferous forest habitats between 7,000 and 10,700 feet elevation. The species annotations include reference to the Merriam Life-zones, names of which appear in the check list as *Lower Sonoran, Upper Sonoran,* etc.

An almost continuous elevated belt of woodland and forest extends across central Arizona. It effectively separates the desert and grassland areas of southern Arizona from the desert and grassland areas of northern Arizona. Accordingly, the desert and grassland animals often exhibit geographic variation within those species which occur both in southern and in northern parts of the state. That is, species which occur on both sides of the central Arizona highlands are often characterized by having different subspecies in the cooler north and in the warmer south. Similarly, different subspecies of animals are often found in the deserts of western Arizona and in the grasslands of eastern Arizona.

Some of the amphibians and reptiles of the state exhibit such geographic variation between north and south and between east and west.

While it is not the purpose of the present series of check lists to treat the subspecies, it is of interest that in Arizona there are subspecies of amphibians and reptiles that are as distinctive in appearance from one another as some species are, and occasionally even more so.

The common names here employed for the species are those published by the Committee on Herpetological Common Names of the American Society of Ichthyologists and Herpetologists (Conant, *et al.*, Copeia, 1956, 3:172-85). However, as results of recent changes in scientific names, new additions to the fauna, or for other reasons, there are a few common names in this check list which will not be found in the "Copeia list."

During the course of recent investigations of the herpetofauna of Arizona, additional data pertinent to the evolutionary relationships of many forms have been accumulated. In fact, many nomenclatorial changes have become necessary for a properly meaningful check list. The writer feels obliged to incorporate some of them in the present list, although the reasons for them — which would (and did originally) comprise an unduly long and somewhat inappropriate taxonomic section here — will appear in another paper to be published elsewhere.

SALAMANDERS

Family Ambystomidae: Ambystomid Salamanders

1. *Ambystoma tigrinum* Green. Tiger Salamander. This species remains the only salamander known to occur in Arizona. It is found within the Colorado River basin and primarily in the northern half of the State, east of Mohave County; a few scattered native populations occur in the southeastern corner (Cochise and Santa Cruz counties). *Upper Sonoran, Transition,* and *Boreal.* Today this species can be collected at localities in Arizona from the desert-grassland upward into spruce-fir forest, over an elevational range from a little under 3,000 feet to a little over 9,000 feet. During dry periods the adults remain well underground and usually in the burrows of other animals such as ground squirrels, kangaroo rats, gophers, etc. It is a particularly hardy species which breeds successfully in both permanent and temporary pools, ponds, lakes, and streams, and in muddy as well as in clear water. The tiger salamander is caught (also reared) in Arizona, and other western states, and sold for fish bait. This practice has introduced and widely mixed several subspecies throughout the warm-water fishery in Arizona.

FROGS AND TOADS

Family Pelobatidae: Spadefoot Toads

2. *Scaphiopus couchi* Baird. Couch's Spadefoot. Southern half of the state; abundant in the southeastern quarter and less frequent in the southwestern quarter where it occurs in the vicinity of Yuma. The northern known limit in the state is the vicinity of Petrified Forest National Monument. *Lower Sonoran* and *Upper Sonoran.* This species lives in a wide variety of desert and grassland habitats where it is strictly a ground dweller — as are all the spadefoot toads. It frequents rocky canyons and bajadas of desert ranges as well as open plains and valley bottomlands, breeding in paloverde, creosotebush, tarbush, and mesquite communities as well as in the desert-grassland and plains grassland.

3. *Scaphiopus (Spea) bombifrons* Cope. Plains Spadefoot. Grasslands primarily in the southeastern corner of the state where it is essentially restricted to plains and valley bottomlands; it occurs northward to the vicinity of Petrified Forest National Monument. *Upper Sonoran.* This species is abundant in Arizona in desert-grassland and plains grassland habitats, where it breeds in the muddy waters of temporary, summer rain-formed pools, ponds, and small lakes — as do all of our spadefoot toads.

4. *Scaphiopus (Spea) intermontanus* Cope. Great Basin Spadefoot. The southern edge of the range of this species extends across the extreme northern part of the state, north of an approximate line Lake Mead-Grand Canyon-Chinle. *Lower Sonoran* upward into *Hudsonian.* While commonest in the high desert region of the Great Basin, in the plains grassland, and in juniper-pinyon woodland, it extends upward into spruce-fir forest in Arizona and Utah. It breeds in summer water that is often much colder than is ordinarily met in Arizona by its close relatives, *S. bombifrons* and *S. hammondi.*

5. *Scaphiopus (Spea) hammondi* Baird. Western Spadefoot. Widely distributed in the eastern half of the state, where its range overlaps that of *S. bombifrons* in the southeastern corner but does not overlap that of *S. intermontanus* in the north. *Lower Sonoran, Upper Sonoran,* and *Transition.* The western spadefoot makes a successful living in a wide variety of habitats from desert and grassland upward through chaparral and woodland into pine forest.

Family Leptodactylidae: Leptodactylid Frogs

6. *Eleutherodactylus augusti* Cope. Barking Frog. Santa Rita and Pajarito Mountains in Santa Cruz County. *Upper Sonoran.* During the summer rainfall period, in July and August, this species has been seen and heard on rocky hillsides of canyons in woodland vegetation, but rarely collected in Arizona.

Family Bufonidae: Toads

7. *Bufo alvarius* Girard. Colorado River Toad. Southern part of the state in the Colorado River basin, Yaqui River basin in the extreme southeastern corner, and basin of Rio de la Concepcion in Santa Cruz County. *Lower Sonoran* into *Upper Sonoran.* Primarily in desert habitats, and less frequent in grassland and the lower edge of oak woodland. Widely distributed on highly varied soils; refuge is commonly taken underground in rodent burrows, where it also "hibernates."

8. *Bufo woodhousei* Girard. Woodhouse's Toad. Eastern half of state, and sparingly westward to the Colorado River. Primarily *Lower Sonoran* into *Upper Sonoran.* A riparian species occurring principally along courses of rivers and permanent and semi-permanent streams (also irrigation ditches, *et al.*), and rarely found at any great distance from their channels or floodplains. It is occasionally found at permanent lakes and ponds.

9. *Bufo microscaphus* Cope. Southwestern Toad. Southeastern quarter of the state. *Lower Sonoran* and *Upper Sonoran* (woodland). Rocky stream canyons and floodplains in the Arizona Upland desert and evergreen woodland south of the Mogollon Rim. Relatively little is known about the bahavior and microenvironmental preferences of this species.

10. *Bufo cognatus* Say. Great Plains Toad. Widespread throughout the state in grasslands and desert. *Lower Sonoran* and *Upper Sonoran* (grassland). While occasionally in low desert ranges this species is absent from mountainous country. It is not dependent on permanent surface waters for reproduction. However, regularly irrigated fields often produce very large local populations, for shorter or longer periods of time, which populations crash upon farming abandonment of the crop fields and subsequent lack of irrigation water.

11. *Bufo punctatus* Baird and Girard. Red-spotted Toad. Throughout the state at widely scattered localities. *Lower Sonoran* and *Upper Sonoran.* A rock dweller in rocky canyons and foothills with or without permanent surface water. Absent from mountainous areas above woodland.

12. *Bufo debilis* Girard. Green Toad. Grasslands in the southeastern corner of the state. *Upper Sonoran.* Extremely common at many localities in Cochise County where spadefoot toads are also abundant during the summer rains.

13. *Bufo retiformis* Smith and Sanders. Sonora Green Toad. Restricted to an extreme southcentral area of the state, near the international line, from the Baboquivari Mountain area westward into the Organ Pipe Cactus National Monument area in Pima County. Primarily *Lower Sonoran.* Often locally abundant in populations which are located in desert and relictual desert-grassland habitats in the Sonoran Desert. Here they congregate at rain-formed pools and ponds also commonly used for breeding by *Pternohyla fodiens* and *Gastrophryne olivacea.*

Family Hylidae: Hylid Frogs

14. *Pternohyla fodiens* Boulenger. Burrowing Casque-head Frog. Restricted to an extreme southcentral area of the state, near the Arizona-Sonora boundary line, between the Baboquivari Mountains on the east and the Organ Pipe Cactus National Monument area on the west. Primarily *Lower Sonoran.* On the ground and burrowing underground during late summer. Often locally abundant in relictual desert-grassland habitats within the Sonoran Desert where they congregate at summer rain-formed pools and ponds.

15. *Hyla wrightorum* Taylor. Arizona Treefrog. Forested central plateau of the state north of the Mogollon Rim; south of the Mogollon Rim this species is as yet known only from the Huachuca Mountains; it is expected to occur also in other forested high mountain ranges. Primarily *Transition,* occasionally *Upper Sonoran* and *Canadian.* On the ground or in shrubs and trees near ponds, pools and streams, usually in coniferous forest.

16. *Hyla arenicolor* Cope. Canyon Treefrog. Widely distributed over most of the state, but absent or rare in the southwestern corner (Yuma County). *Lower Sonoran* and *Upper Sonoran* into *Transition.* This is a riparian species occurring almost exclusively in boulder-strewn canyons with permanent or essentially permanent streams, from desert to ponderosa pine forest at 8,000-8,500 feet elevation. Often there is a variable canpoy of broadleaf deciduous trees such as cottonwoods, willows, sycamores, ashes, walnuts, mulberrys, locusts, chokecherrys, alders, etc. Populations of this species are remarkably camouflage-adapted in color and pattern harmony with the streamside rocks and boulders on which the animals usually sit.

17. *Pseudacris nigrita* Le Conte. Chorus Frog. High grasslands and forests of the central plateau area of the state. *Upper Sonoran, Transition,* and lower parts of *Canadian.* This species frequents meadows, lake margins, and generally marshy habitats where it is usually found on the ground or in low plants near the ground.

Family Microhylidae: Microhylids

18. *Gastrophryne carolinensis* Holbr. Woodland Narrow-mouthed Toad. Known from a few localities only in the Pajarito and Patagonia Mountains in the extreme southcentral part of the state. *Upper Sonoran.* During the summer rains this species is found in and near rain-formed ponds and pools, as well as in permanent clear water streams, in oak woodland and oak-grass habitats. Here *Rana pipiens* and/or *Scaphiopus hammondi* are locally plentiful, *Rana tarahumarae* and/or *Bufo punctatus* usually present, and *Bufo alvarius* may be present but is much less frequent than the others.

19. *Gastrophryne olivacea* Hallow. Plains Narrow-mouthed Toad. Known from a few localities in Pima County. *Lower Sonoran.* The plains narrow-mouthed toad recently has been found in and near summer rain-formed ponds and pools on the Papago Indian Reservation west and northwest of Sells, eastward to Three-Points, 20 miles southwest of Tucson.

It lives in desert habitats and in relictual mesquite-grass habitats within the Sonoran Desert, where it shares the mud-brown breeding waters with such abundant species as *Pternohyla fodiens, Bufo retiformis, Bufo alvarius, Bufo cognatus, Scaphiopus couchi,* and others, including occasional individuals of *Rana pipiens* and *Kinosternon flavescens.*

Family Ranidae: Frogs

20. *Rana catesbeiana* Shaw. Bullfrog. Introduced. The eastern bullfrog has been widely introduced into Arizona, and is established at numerous localities where permanent water occurs; these localities are principally in the southeastern quarter of the state. *Lower Sonoran* and *Upper Sonoran.* This species, the largest amphibian in the state's fauna, inhabits lakes, permanent ponds, and occasionally smaller pools or ponds regularly recharged by run-off from summer rains. It is also successful in permanent rivers and streams.

21. *Rana tarahumarae* Boulenger. Tarahumara Frog. Known in Arizona (and United States) only from the Pajarito Mountains in Santa Cruz County, where it has been taken in Sycamore Canyon, Pena Blanca Canyon, and Alamo Canyon. *Upper Sonoran* (woodland). This frog inhabits permanent streams and springs, and is rarely found at a pond or pool very far from a spring or stream channel.

22. *Rana pipiens* Schreber. Leopard Frog. Numerous isolated populations of this species occur almost throughout the state at localities where permanent or semi-permanent surface waters occur. Primarily *Lower Sonoran* and *Upper Sonoran;* also *Transition.* This widespread and highly successful frog is often common or abundant in springs, streams, rivers, lakes, ponds, cattle tanks, etc., from the desert upward into pine forest. Marked geographic variation is exhibited by distinctive populations of this species in Arizona.

TURTLES

Family Kinosternidae: Musk and Mud Turtles

1. *Kinosternon flavescens* Agassiz. Yellow Mud Turtle. Southeastern part of the state, in permanent or temporary waters in the desert-grasslands in Cochise and Pima counties. Primarily *Upper Sonoran.* Semi-aquatic, in streams as well as in temporary rain-formed pools and ponds which result from the southwestern summer monsoon which occurs from late June into September. This mud turtle is less abundant in Arizona than is the following species.

2. *Kinosternon sonoriense* Le Conte. Sonoran Mud Turtle. Southcentral and southeastern area of the state. *Lower Sonoran* and *Upper Sonoran.* Semi-aquatic, in streams both permanent and semi-permanent. A common species in streams in the Arizona Upland desert and in oak woodland.

Family Emydidae: Fresh-water and Marsh Turtles

3. *Terrapene ornata* Agassiz. Western Box Turtle. Grassland in the southeastern corner of the state, in Cochise and Pima counties. *Upper Sonoran;* marginal in *Lower Sonoran.* While primarily a terrestrial species, the western box turtle often enters rain-formed pools and ponds during the summer. It is particularly abundant at many localities in the grasslands of Cochise County.

Family Testudinidae: Land Tortoises

4. *Gopherus agassizi* Cooper. Desert Tortoise. Sonoran and Mohave deserts in the southern, western, and extreme northwestern parts of the state. *Lower Sonoran;* marginal in *Upper Sonoran.* The desert tortoise is strictly terrestrial and a vegetarian. While nowhere very abundant, it remains locally a common species yet actually little studied and little known in its natural home. In Arizona and Sonora it is found predominantly in rocky foothills where it digs burrows into hillsides and under rocks, often taking advantage of initial excavations by other animals, particularly the rock squirrel (*Citellus variegatus*) and the Harris antelope squirrel (*C. harrisi*).

Family Trionychidae: Softshell Turtles

5. *Trionyx spinifera* Schneider. Spiny Softshell. Introduced. This species was introduced, via Gila River in New Mexico, into the Gila and Colorado River basins in the southern and western parts of the state. *Lower Sonoran* and *Upper Sonoran.* The softshell is the most strictly aquatic of our turtles and is found in permanent rivers and streams. Occasionally it gets into other situations, such as irrigation canals, where it sometimes becomes stranded.

LIZARDS

Family Helodermatidae: Beaded Lizards

6. *Heloderma suspectum* Cope. Gila Monster. Southern half into the extreme northwestern area of the state; primarily in the Sonoran Desert and extreme western edge of the Mohave Desert, but also occurs less frequently in the desert-grassland and rarely into oak woodland to 4,100 feet. *Lower Sonoran* into *Upper Sonoran.* Principally in undulating rocky foothill terrain, on rocky-gravelly bajadas, and in rocky canyons; less frequent or absent on open sandy desert plains, in the desert-grassland and lower edge of evergreen woodland. Shelter is taken in rodent burrows, wood rat (pack rat) nests, dense thickets and under rocks; winter denning occasionally occurs among rocks similar to and near dens of the western diamondback rattlesnake (*Crotalus atrox*).

Family Gekkonidae: Geckos

7. *Coleonyx variegatus* Baird. Banded Gecko. Widely distributed throughout the Sonoran and Mohave deserts, south and west of the Mogollon Rim; occasionally in desert-grassland. *Lower Sonoran* into *Upper Sonoran.* Common on sandy and gravelly soils with rodent burrows used for shelter, and also frequent in rocky terrain; often climbs on rock surfaces and in crevices.

Family Iguanidae: Iguanas, *et al.*

8. *Dipsosaurus dorsalis* Baird and Girard. Desert Iguana. Widely distributed in the southwestern area of the state, in the Sonoran and Mohave deserts to 3,300 feet; in extreme southern Arizona the range extends eastward to the Tucson Mountains. *Lower Sonoran.* Common to abundant in valleys and plains where creosotebushes (*Larrea divaricata*), with rodent holes and burrows under them, are characteristically present.

9. *Crotaphytus collaris* Say. Collared Lizard. Desert and grassland into evergreen woodland almost throughout the state. *Lower Sonoran* and *Upper Sonoran.* Rocks, large or small, single or numerous, are used for immediate shelter and may occur in open flat terrain as well as on rocky hills and mountain slopes.

10. *Gambelia wislizeni* Baird and Girard. Leopard Lizard. Desert and grassland almost throughout the state. *Lower Sonoran* and *Upper Sonoran.* A swift ground dweller and lizard eater that is nowhere a relatively abundant species, but one which occurs at least sparingly in almost every type of arid and semi-arid terrain throughout the deserts and grasslands; from sand to rock and from essentially bare to densely covered shrub and tree habitats.

11. *Sauromalus obesus* Baird. Chuckwalla. Widely distributed in the Sonoran and Mohave deserts, south and west of the Mogollon Rim and in the Grand Canyon along the Colorado River into extreme southern Utah; eastward in extreme southern Arizona to the Silverbell Mountains in Pima County. *Lower Sonoran.* Rocks and rocky crevices are used for night and day shelter, sunning stations and hibernation; scattered rocks are used for temporary shelter during diurnal foraging away from home crevices.

12. *Holbrookia maculata* Girard. Lesser Earless Lizard. In the eastern half of the state. *Upper Sonoran.* A grassland species on the ground and in rodent holes on rockless plains and bottom lands; also on rocky hillsides and bajadas where scattered rocks are commonly used for resting and for shelter.

13. *Holbrookia elegans* Bocourt. Madrean Earless Lizard. Southcentral area of the state in the Sonoran Desert westward to Ventana Ranch in Pima County, and in the oak woodland eastward to the Huachuca Mountains in Cochise County; southward in desertscrub and into thornscrub in southern Sonora and Sinaloa. *Lower Sonoran* and *Upper Sonoran.* Frequents rocky foothills and slopes in the Arizona Upland desert between 1,000 and 4,000 feet elevation, and in oak-grass habitats above 4,000 feet.

14. *Holbrookia texana* Troschel. Greater Earless Lizard. Upper edge of the desert into lower evergreen woodland, and less frequently in grassland; principally in the southeastern area of the state with a westward extension to the eastern edge of the Mohave Desert at the Williams River. *Upper Sonoran* and *Lower Sonoran.* On the ground and on small rocks in rocky canyons, on rocky hillsides, and on rocky bajadas at the bases of desert ranges.

15. *Callisaurus draconoides* Blainville. Zebra-tailed Lizard. Widely distributed in the Sonoran and Mohave deserts in the southwestern area, extending eastward into the Chihuahuan-Sonoran Desert transition in Cochise County. *Lower Sonoran.* Open, flat sandy desert valleys and plains, with particular abundance in washes and arroyos and along the margins of dunes; widely occurring over most arid flat terrain but conspicuously less abundant on rock and gravel surfaces.

16. *Uma notata* Baird. Fringe-toed Lizard. (Note: The species *notata* here includes the form *scoparia.*) Extreme southwestern corner of the state in Yuma County. *Lower Sonoran.* Restricted to fine wind-blown sand dunes and similar loose sand accumulations in dune areas.

17. *Sceloporus scalaris* Wiegmann. Bunch Grass Lizard. Sierra Madrean ranges in the southeastern area of the state; primarily coniferous forest in the Santa Rita, Dragoon, Huachuca and Chiricahua Mountains mainly above 6,000 feet; recently taken in desert-grassland at 4,300 feet. *Transition* and *Boreal,* infrequent or rare in *Upper Sonoran.* On the ground in and among bunchgrass, usually in open sunned patches on south-facing slopes in coniferous forest.

18. *Sceloporus jarrovi* Cope. Mountain Spiny Lizard. Woodland and coniferous forest in the Baboquivari, Santa Rita, Dragoon, Huachuca, Chiricahua, and Pinaleno (Graham) mountains, primarily above 5,000 feet; occurs to as low as 4,800 feet and as high as 10,700 (Mt. Graham). *Upper Sonoran, Transition,* and *Boreal.* Principally in rocky habitats; common on rock cliffs and talus slopes of hillsides which are usually south, west or east-facing; less frequent or absent in canyon bottoms.

19. *Sceloporus undulatus* Latreille. Eastern Fence Lizard. Eastern and northern parts of the state. On the Colorado Plateau, southward on forested mountains to the Santa Catalinas in Pima County, and in the desert-grassland. Primarily in *Upper Sonoran* and *Transition;* occasionally marginal in *Lower Sonoran,* as at Safford, Graham County. Variable terrain from flat sandy grasslands to areas of rock in coniferous forest; climbs and suns on yuccas, trees, logs, rocks, fenceposts, etc.

20. *Sceloporus virgatus* Smith. Striped Plateau Lizard. Chiricahua Mountains (southward in Sierra Madrean ranges) in woodland and coniferous forest from 4,900 feet to 8,400 feet. *Upper Sonoran* and *Transition.* Principally on rocks and boulders, also on logs, etc., in canyon bottoms where often commonest along the banks and in the immediate vicinity of the drainage channels; usually less frequent or absent on rocks, talus, etc., high on canyon hillsides.

21. *Sceloporus clarki* Baird and Girard. Sonora Spiny Lizard. Southeastern quarter of the state, with a westernmost disjunctive population in the Ajo Mountains in western Pima County. Primarily *Upper Sonoran* (evergreen woodland), marginal in *Lower Sonoran* with local abundance in well-developed riparian woodlands along major drainageways in the Sonoran Desert. On trees and their adjacent rocks, and occasionally in rocky habitats without trees at the upper edge of the desert and in the desert-grassland; greatest abundance is on and among evergreen and deciduous trees in woodland communities.

22. *Sceloporus magister* Hallowell. Desert Spiny Lizard. Widely distributed in the Sonoran and Mohave deserts and in the desert-grassland. *Lower Sonoran* and *Upper Sonoran.* Primarily on the ground, also on rocks, taking refuge in dense clumps of vegetation, in rodent burrows and under rocks; in areas where there is scant rock or shrub cover, trees (e.g., Joshua tree, *Yucca brevifolia*) may be used for shelter and sunning, with hibernation a few inches underground in sandy soils.

23. *Sceloporus graciosus* Baird and Girard. Sagebrush Lizard. Across the extreme northern part of the state. *Lower Sonoran* (Great Basin Desert), *Upper Sonoran* (woodland), *Transition,* and *Canadian.* Primarily on the ground in basin sagebrush (*Artemisia tridentata*) and other shrub cover (occasionally rocks and litter) in the open desert, and in similar shrubby openings in juniper-pinyon woodland and coniferous forest.

24. *Urosaurus graciosus* Hallowell. Long-tailed Brush Lizard. Sonoran and Mohave deserts; eastward in the extreme southern part of the state to the Avra Valley (west of Tucson Mountains); northwestward to Boulder Dam. *Lower Sonoran.* A climbing species on desert trees (e.g., paloverde) and desert shrubs (e.g., creosotebush) that is rarely seen on the ground or on rocks.

25. *Urosaurus ornatus* Baird and Girard. Tree Lizard. Widely distributed throughout the state from approximately 100 feet on the banks of the Colorado River in the Yuma region into coniferous forest in several mountains. *Lower Sonoran, Upper Sonoran,* and *Transition.* A climbing species on small-leaf and broadleaf trees and shrubs, and on rocks and cliffs, from open desert to sunned areas in woodland and forest; rarely seen more than momentarily on the ground, and then usually running, although commonly found sitting or climbing on large rocks and boulders.

26. ***Uta stansburiana*** Baird and Girard. Side-blotched Lizard. Widely distributed throughout the state. *Lower Sonoran* and *Upper Sonoran.* A ground species in open, flat sandy desert and grassland terrain with shelter usually under shrubs, in rodent holes and in wood rat nests; also on the ground and rocks on bajadas, rocky hillsides etc., with shelter usually under rocks.

27. ***Phrynosoma douglassi*** Bell. Short-horned Lizard. Essentially in the eastern half of the state from approximately 4,200 to 10,400 feet; not a desert species. *Upper Sonoran, Transition,* and *Boreal.* Flat, semi-arid grassland plains to mountain hillsides and valleys in spruce-fir forest.

28. ***Phrynosoma cornutum*** Harlan. Texas Horned Lizard. Chihuahuan Desert and desert-grassland in the extreme southeastern corner of the state in Cochise County. *Lower Sonoran* and *Upper Sonoran* (grassland). Sandy or gravelly flat ground with or without rocks, usually with scattered desert and grassland shrubs; often in the same habitat with *P. modestum.*

29. ***Phrynosoma modestum*** Girard. Round-tailed Horned Lizard. Chihuahuan Desert and desert-grassland in the extreme southeastern corner of the state in Cochise and Graham counties. *Lower Sonoran* and *Upper Sonoran* (grassland). Sandy or gravelly soil with or without rocks, usually with scattered desert shrubs; often in the same habitat with *P. cornutum.*

30. ***Phrynosoma solare*** Gray. Regal Horned Lizard. Southcentral area of the Arizona Upland region of the Sonoran Desert, and along the adjacent western edge of the desert-grassland. *Lower Sonoran* into *Upper Sonoran.* Rocky and gravelly bajadas and desert foothills; less frequent or absent on sandy plains and rarely in the same habitat with any other species of horned lizard (occasional in southeastern Cochise County).

31. ***Phrynosoma platyrhinos*** Girard. Desert Horned Lizard. Sonoran and Mohave Deserts in western Arizona extending eastward across the extreme southern half of the state to the Avra Valley (west of Tucson). *Lower Sonoran.* Occurs on a wide range of desert soils from loose fine sand to substrata quite gravelly and rocky in texture; usually found in the vicinity of considerable shrub cover as compared to the next species listed with which it is occasionally sympatric.

32. ***Phrynosoma m'calli*** Hallowell. Flat-tailed Horned Lizard. Extreme southwestern corner of the state, in Yuma County, which is a part of its relatively restricted range within the Sonoran Desert, i.e., the Lower Colorado four-corners of California-Baja California-Sonora-Arizona. *Lower Sonoran.* Fine sandy soils, not gravelly nor rocky, with or without scant vegetation; occurs commonly with *Uma notata* on lower parts of loose wind-blown dunes.

Family Xantusidae: Night Lizards

33. *Xantusia arizonae* Klauber. Arizona Night Lizard. Restricted to upper desert and lower woodland habitats along the southwestern edge of the Colorado Plateau in the west central part of the state, from the vicinity of Valentine, Mohave County, southeastward to the Superstition Mountains, Maricopa County. *Upper Sonoran* into *Lower Sonoran.* Rocks and rocky hillsides in chaparral and in areas with mixed desert and woodland vegetation; under cap flakes of massive granite boulders, in rock crevices, and occasionally under yucca logs and clumps in the vicinity of boulders.

34. *Xantusia vigilis* Baird. Desert Night Lizard. Mohave Desert in the northwestern corner of the state, and in the Kofa Mountains in Yuma County (Sonoran Desert). *Lower Sonoran.* Under and in dead and rotting branches and leaves, trunks and stumps of the Joshua tree (*Yucca brevifolia*) and, occasionally, other yuccas; occasionally under litter in and along the edge of Joshua tree "forest."

Family Scincidae: Skinks

35. *Eumeces callicephalus* Bocourt. Mountain Skink. Sierra Madrean ranges in the southeastern corner of the state; Baboquivari, Pajarito, Santa Rita, and Huachuca mountains (has been reported with question from the Chiricahua Mountains) from approximately 4,000 to 6,500 feet. *Upper Sonoran* and *Transition.* Under and among rocks, rock fragments, and logs in canyon bottoms and on hillsides in woodland and forest.

36. *Eumeces multivirgatus* Hallowell. Many-lined Skink. Northern part of the state primarily on the Colorado Plateau, where it occurs to 7,200 feet. Primarily *Upper Sonoran* and *Transition.* Under rocks, logs, and litter on plains, and in canyon bottoms and on hillsides in woodland and forest.

37. *Eumeces obsoletus* Baird and Girard. Great Plains Skink. South-central and southeastern Arizona, from approximately 3,000 to 7,000 feet. Primarily *Upper Sonoran* and *Transition.* Under large boulders and rocks, also in burrows and under litter; principally on the banks or in the near vicinity of permanent and semi-permanent streams or arroyos in the desert, grassland, and mountains. In the desert it is a strictly riparian species, and it has not been taken west of the Santa Catalina Mountains, near Tucson.

38. *Eumeces gilberti* Van Denburgh. Gilbert's Skink. Evergreen woodland in the Bradshaw Mountains, and riparian woodland along the Hassayampa River in Yavapai and Maricopa counties. *Upper Sonoran* into *Lower Sonoran* (riparian). Among rocks, logs, and leaf litter under the canopy of broadleaf deciduous riparian trees (cottonwood-willow association) near the banks of permanent and semi-permanent streams in the desert, chaparral, and woodland.

39. *Eumeces skiltonianus* Baird and Girard. Western Skink. Known only from the Kaibab Plateau, Coconino County, where it has been taken in the Bright Angel Creek area below the North Rim of Grand Canyon, and at Jacobs Lake. *Transition* and *Canadian.* Under logs and rocks in openings of ponderosa pine forest and fir forest habitats, and in rocky streamside habitats under canopies of broadleaf deciduous trees.

Family Teidae: Teids

40. *Cnemidophorus burti* Taylor. Sonora Whiptail. In the Santa Cruz and San Pedro River basins of the southcentral area of the state, from the Santa Catalina Mountains (vicinity of Oracle) southward; also in Guadalupe Canyon in the extreme southeast corner of the state (Yaqui River basin), and in the basin of Rio de la Concepcion in Santa Cruz County, isolated in desert mountains westward to the Ajo Mountains. *Lower Sonoran* (primarily riparian) and *Upper Sonoran.* Among dense shrubby vegetation, and often rocks, near and on the banks of permanent and semi-permanent streams and arroyos in the desert, desert-grassland, and evergreen woodland to approximately 4,500 feet.

41. *Cnemidophorus exsanguis* Lowe. Spotted Whiptail. Southeastern Arizona principally in evergreen woodland and in ponderosa pine forest; also at lower elevations at scattered localities along the upper edge of the desert and desert-grassland where it may be locally abundant in rocky or riparian situations. Primarily *Upper Sonoran,* into both *Transition* and *Lower Sonoran.* This is an "all-female" species.

42. *Cnemidophorus velox* Springer. Plateau Whiptail. Colorado Plateau of the central and northern parts of the state, and southward on forested mountains as far south as Pinal Peak in Gila County, where it occurs to *ca.* 8,000 feet. *Upper Sonoran* (woodland) and *Transition.* Open, sunned areas and rocky sites in semi-arid juniper-pinyon woodland, ponderosa pine forest, and at somewhat lower elevations in broadleaf riparian woodland in the vicinity of permanent and semi-permanent streams. This is an "all-female" species.

43. *Cnemidophorus arizonae* Van Denburgh. Arizona Whiptail. Grassland and Chihuahuan Desert habitats in the southeastern area of the state; also in evergreen woodland to approximately 5,000 feet. *Upper Sonoran* and *Lower Sonoran.* Predominantly on plains and gentle foothill slopes, occasionally in areas with scant cover (grasses and herbs with or without a few scattered shrubs), but more commonly where mesquite and yucca are present, and often abundant where mesquite is dense on much-deteriorated grassland. This is an "all-female" species.

44. *Cnemidophorus inornatus* Baird. Little Striped Whiptail. Grassland in Cochise County where its range overlaps that of the superficially similar Arizona Whiptail *(C. arizonae).* This species occurs in Arizona as a distinctive, geographically isolated population known only from the vicinity of the Willcox Playa. *Upper Sonoran.* In open, short-grass habitats in the desert-grassland with little or no yucca present and virtually no mesquite compared to the dense mesquite thickets commonly inhabited by *C. arizonae;* summer annuals are usually present and often abundant. Small rodent holes are used for shelter, as is usual for whiptails.

45. ***Cnemidophorus tigris*** Baird and Girard. Western Whiptail. Widely distributed in the state in desert, grassland, and evergreen woodland below approximately 4,500 feet. *Lower Sonoran* and *Upper Sonoran.* Present over highly variable terrain from arid sandy plains and rock desert landscapes into semi-arid grasslands and juniper-pinyon woodland.

Family Anguidae: Anguids

46. ***Gerrhonotus kingi*** Gray. Arizona Alligator Lizard. Evergreen woodland and coniferous forest in the southeastern area of the state below 7,500 feet. Also at lower elevations (to 2,400 feet) in broadleaf riparian woodland along major drainageways in desert and grassland. *Transition, Upper Sonoran, Lower Sonoran* (riparian). Principally in canyons and on hillsides near and under logs of oak, pine, cottonwood and other trees, under rocks and litter, and occasionally in wood rat nests and rodent burrows.

SNAKES

Family Leptotyphlopidae: Blind Snakes (or Worm Snakes)

47. ***Leptotyphlops dulcis*** Baird and Girard. Texas Blind Snake; Texas Worm Snake. Grassland and Chihuahuan Desert in the extreme southeastern corner of the state. *Upper Sonoran* and *Lower Sonoran.* Subterranean burrower (fossorial) in partially sandy and loamy semiarid grassland plains and the upper edge of the desert; among roots of trees and shrubs, under rocks and in rock crevices, and under wood and litter on the surface of the ground.

48. ***Leptotyphlops humilis*** Baird and Girard. Western Blind Snake; Western Worm Snake. Widespread in desert and grassland in southern and extreme northwestern parts of the state, less frequent in woodland. Primarily *Lower Sonoran;* in the Mohave, Sonoran, and Chihuahuan deserts. Fossorial in sandy, loamy, and occasionally gravelly and stony soil; under rocks and in rock crevices, among roots of desert trees and shrubs, and under ground litter.

Family Boidae: Boas

49. ***Lichanura trivirgata*** Cope. Desert Boa (subspecies are either chocolate-striped or reddish-striped "rosy boas"). In the Sonoran Desert, and southeastern edge of the Mohave Desert, eastward in southern Arizona to the Harquahala Mountains and the vicinity of Aguila, Maricopa County; and southward through the Organ Pipe Cactus National Monument area into Sonora, Mexico. *Lower Sonoran.* It is crepuscular and nocturnal, in rocky areas in desert ranges; also rarely encountered in open rockless terrain.

Family Colubridae: Colubrids

50. *Natrix rufipunctatus* Cope. Narrow-headed Water Snake. East central part of the state from the Colorado Plateau (vicinity of Flagstaff) southeastward to headwaters of the Gila River in Greenlee County. *Upper Sonoran* and *Transition.* Almost strictly aquatic; in quiet, often deep and rocky pools along permanently flowing streams (such as Oak Creek and Eagle Creek) commonly sheltered by broadleaf deciduous trees (cottonwood-willow association) in juniper-pinyon and oak woodland, into ponderosa pine forest.

51. *Thamnophis eques* Reuss. Mexican Garter Snake. Southeastern quarter of state. Known from several scattered valley and bottomland localities in the grassland, occasionally in the desert, and in lower parts of oak woodland. *Upper* and *Lower Sonoran.* In or near water along streams coursing valley floors and generally open areas; not in mountain canyon stream habitats.

52. *Thamnophis cyrtopsis* Kennicott. Black-necked Garter Snake. Throughout most of the eastern half of the state, disjunctive westward to the Ajo Mountains in western Pima County. *Lower Sonoran, Upper Sonoran,* into *Transition.* In and along the banks of mountain canyon streams with intermittent or permanent flow in desert, grassland, woodland, and occasionally in lower ponderosa pine forest, to 6,800 feet.

53. *Thamnophis marcianus* Baird and Girard. Checkered Garter Snake. Widespread over plains and valleys across much of the southern half of the state in the desert and desert-grassland. *Upper Sonoran* and *Lower Sonoran.* Often found several miles distant from, as well as near to, rivers, streams and ponds in arid and semi-arid habitats; rodent burrows are often used for shelter. In the desert, summer feeding and general activity are primarily at night, particularly in situations distant from water.

54. *Thamnophis elegans* Baird and Girard. Western Garter Snake. Widespread over the northeastern and extreme northwestern parts of the state in grassland, woodland, and forest to over 9,000 feet. *Upper Sonoran* and *Transition.* Frequents permanent and intermittent streams and ponds, and often occurs at considerable distances from surface water.

55. *Heterodon nascius* Baird and Girard. Western Hognose Snake. Grassland in the southeastern part of the state. Primarily *Upper Sonoran.* Principally in open, flat or rolling plains in the grassland; occasionally into mountain canyon bottoms or flood plains of streams with canopies of broadleaf deciduous trees (sycamore, walnut, cottonwood, willow, etc.) in evergreen woodland to 5,100 feet.

56. *Masticophis bilineatus* Jan. Sonora Whipsnake. Southern half of the state; not in the desert in Yuma County. Principally *Upper Sonoran,* into *Lower Sonoran.* Evergreen woodland to 6,100 feet and occasional in grassland; in the Arizona Upland desert (paloverde-sahuaro association) to as low as 2,600 feet in the Ajo and Tucson Mountains, in Pima County. Shelter and foraging on ground, in shrubs and in trees.

57. *Masticophis flagellum* Shaw. Coachwhip; Whipsnake (Red Racer; Black Racer). (Note: Color phases within the single subspecies *M. f. piceus* in Arizona and elsewhere are called red racer and black racer; there are also intermediate brown, gray and variously speckled morphs.) Widely distributed over the deserts and grasslands of the state. Principally *Lower Sonoran* and *Upper Sonoran.* Rapid forager in highly varied terrain (to approximately 4,500 feet), that may be flat or hilly, sandy or rocky; and grass, shrub or tree dominated. Shelter in rodent burrows, wood rat nests, under and among rocks; commonly ascends shrubs and trees for escape as well as for forage.

58. *Masticophis taeniatus* Hallowell. Striped Whipsnake. Widely distributed over most of the northern half of the state, along the upper edge of the Great Basin Desert, in grassland, juniper-pinyon woodland, and into the lower edge of ponderosa pine forest to approximately 6,000 feet. Primarily *Upper Sonoran.* Open plains and mesas in canyons and along streamways; shelter and foraging in rocky outcrops, rodent burrows, and in trees and shrubs.

59. *Salvadora grahamiae* Baird and Girard. Mountain Patch-nosed Snake. Mountains of the southeastern corner of the state in and above oak-grass habitats at about 4,500 feet. *Upper Sonoran* and *Transition.* Ground dwelling and fast moving, preferring open, sunny and often rocky areas in evergreen woodland and lower ponderosa pine forest.

60. *Salvadora hexalepis* Cope. Desert Patch-nosed Snake. Widely distributed in the southern half and in the northwestern part of the state, in deserts, grasslands, and occasionally into the lower edge of chaparral and woodland. *Lower Sonoran* and *Upper Sonoran.* Ground dwelling and fast moving, preferring open, flat sunny areas where it is commonly found stretched out and sunning itself in the morning and late afternoon; a ground forager which rarely climbs in shrubs.

61. *Elaphe triaspis* Cope. Green Rat Snake. Restricted to Sierra Madrean ranges ("Mexican Mountains") in the southeastern part of the state; Baboquivari, Pajarito, Santa Rita, and Chiricahua Mountains. Primarily broadleaf riparian associations in evergreen woodland at mid-elevation between 4,000 and 6,000 feet. *Upper Sonoran.* It has been encountered crawling on ground beside streams, crossing roads and trails in canyons, and on rocky hillsides; most situations have been at or near rocky areas and with broadleaf trees nearby.

62. ***Diadophis punctatus*** Linnaeus. Ringneck Snake. Principally in the southeastern quarter of the state, and occuring at scattered localities as far to the northwest as the vicinity of Kingman. *Upper Sonoran;* occasionally as a riparian species in *Lower Sonoran.* Primarily a woodland inhabitant, below 5,500 feet, which also occurs in grassland and extends into the desert along major drainageways to as low as 2,400 feet. It is occasionally seen crawling on the surface, but is usually found under rocks, logs, boards, and other surface cover, most often in relatively humid situations, especially when the ground is damp beneath covering objects.

63. ***Pituophis melanoleucus*** Daudin. Bullsnake; Gopher Snake. Statewide from 100 feet elevation at Yuma in the Sonoran Desert to over 9,000 feet on several mountains, and on the North Rim of the Grand Canyon, in spruce-fir forest. *Lower Sonoran* to *Hudsonian.* This common and highly adaptable species has a greater ecological range and wider geographical distribution than any other reptilian species occurring in Arizona. Principally a ground dweller and rodent burrow hunter, it is a capable digger in loose earth and is a good climber and bird nest hunter in trees and shrubs. It is commonly found on the ground surface, often crossing roads during day or night, and under rocks, logs, boards, and trash.

64. ***Arizona elegans*** Kennicott. Glossy Snake. Deserts and grasslands in the western, southern, and east-central parts of the state. *Lower Sonoran* and *Upper Sonoran.* A strong burrower which commonly seeks shelter in rodent burrows and remains underground during daylight hours, rather than under surface covering objects such as rocks and litter.

65. ***Rhinocheilus lecontei*** Baird and Girard. Long-nosed Snake. Deserts and grasslands in the southwestern, southeastern, and extreme northwestern parts of the state. *Lower Sonoran* and *Upper Sonoran.* As the preceding species, this is a strong burrower which commonly seeks shelter in rodent burrows and remains underground during daylight hours, rather than under surface covering objects.

66. ***Lampropeltis getulus*** Linnaeus. Common Kingsnake. Widely distributed throughout the state below 4,500 feet, except in the northeast where it is rare or absent. Principally *Lower Sonoran* and *Upper Sonoran* (grassland). A highly adaptable ground dweller and burrow hunter which also forages in trees and shrubs; notably active in the desert and desert grassland during and immediately after summer rains.

67. ***Lampropeltis doliata*** Linneaus. Milk Snake. Northeastern area of state. Known in Arizona from two specimens collected 30-40 miles northeast of Flagstaff, on Highway 89, in the vicinity of Wupatki National Monument. *Upper Sonoran.* Open, flat and undulating terrain with scant vegetation.

68. *Lampropeltis pyromelana* Cope. Sonora Mountain Kingsnake. Mountain habitats above 4,700 feet (e.g., oak woodland) into ponderosa pine and Douglas fir forest to 9,100 feet; from the Chiricahua Mountains in the extreme southeastern corner of the state northward and westward into southwestern Utah; absent from most of the southwestern part of the state. Usually found crawling on the surface in mountain canyons; occasionally found under logs, rocks and other surface objects, although usually going underground in rodent burrows for cover; also a capable climber in shrubs and trees but rarely seen above the surface of the ground.

69. *Phyllorhynchus browni* Stejneger. Saddled Leaf-nosed Snake. Southcentral area of the state within the Sonoran Desert, between the vicinity of Ajo on the west and Tucson on the east. *Lower Sonoran.* Fossorial in relatively coarse, rocky soils as well as in sand; usually occurs in rocky-gravelly foothills in the Arizona Upland desert (paloverde-sahuaro association) and is absent or less frequent on open sandy plains.

70. *Phyllorhynchus decurtatus* Cope. Spotted Leaf-nosed Snake. Southwestern part of the state in the Sonoran Desert and Mohave Desert, occurring as far east as the Tucson area. *Lower Sonoran.* Fossorial in sandy soil and common on open desert plains; less frequent (relatively rare) where it occasionally occurs with *P. browni* in rocky-gravelly desert foothills.

71. *Ficimia quadrangularis* Günther. Sonora Hooked-nosed Snake. Oak-grass and mesquite-grass habitats in the Patagonia-Pajarito Mountains area in Santa Cruz County. *Upper Sonoran.* All Arizona specimens (5) of this small ground-dweller have been collected during the summer and fall, between 3,400 and 4,400 feet elevation.

72. *Ficimia cana* Cope. Western Hook-nosed Snake. Known from a few localities in the desert and desert-grassland transition in Cochise County. *Lower* and *Upper Sonoran.* Open desert and grassland habitats with sandy to coarse gravelly soils; a small ground dweller which takes refuge, at least occasionally if not characteristically, in rodent burrows.

73. *Sonora semiannulata* Baird and Girard. Western Ground Snake. Known from scattered localities in the desert and grassland in the southern half and the northwestern quarter of the state. *Lower Sonoran* and *Upper* Sonoran. Ground dweller commonly seeking shelter under surface rocks as well as in rodent holes or sandy plains and mesas; also on rocky hillsides.

74. *Chionactis occipitalis* Hallowell. Western Shovel-nosed Snake. Sonoran and Mohave Deserts in the southwestern part of the state. *Lower Sonoran.* Fossorial in sand and absent or infrequent in rocky desert terrain; most abundant in flat and open sparsely vegetated areas with fine wind-blown sand and dunes. A rapid burrower which is able to "swim" under the surface of the sand.

75. *Chionactis palarostris* Klauber. Sonora Shovel-nosed Snake. In Arizona (and United States) only in Organ Pipe Cactus National Monument, in the Sonoran Desert in southwestern Pima County. *Lower Sonoran.* Fossorial in sand and sandy-gravelly soils in the Arizona Upland desert (paloverde-sahuaro association) where it frequents more coarse (stony and rocky) soils or bajadas and hilly terrain than ordinarly frequented by *C. occipitalis.*

76. *Chilomeniscus cinctus* Cope. Banded Burrowing Snake. Extreme southern part of the state; in the Sonoran Desert eastward to the San Pedro River Valley and northeastward to the vicinity of Superior. *Lower Sonoran.* Fossorial in sandy and sandy-gravelly soils, occurring in both open and sandy desert (creosotebush association) and in the sandy-gravelly washes and arroyos in otherwise rocky upland terrain (paloverde-sahuaro association.)

77. *Oxybelis aeneus* Wagler. Vine Snake. In Arizona (and United States) only in the Pajarito Mountains, Santa Cruz County, within five miles of the Arizona-Sonora boundary where it has been but rarely encountered. *Upper Sonoran.* In shrubs and trees along canyon bottoms and on hillsides, where it resembles a grayish vine and is usually difficult to detect; also among grasses and reeds at the sides of streams and ponds.

78. *Trimorphodon lyrophanes* Cope. Southwestern Lyre Snake. Southern half and northwestern quarter of the state, in desert, desert-grassland, evergreen woodland, and ponderosa pine forest to 7,400 feet. Principally *Lower Sonoran* (Sonoran and Mohave Deserts) into *Upper Sonoran* and *Transition.* A strong and agile climber among rocks and vegetation, primarily in rocky canyons and on rocky hillsides; individuals often climb to 15 or more feet in trees and in rock crevices both of which are used for diurnal and winter shelter as well as for foraging. Rarely encountered in open, rockless or treeless terrain.

79. *Hypsiglena torquata* Günther. Night Snake. Statewide in desert, grassland, and woodland. *Lower Sonoran* and *Upper Sonoran* to 6,400 feet. A ground dweller with wide habitat tolerance as indicated by its occurrence on desert mountains with shelter under rocks, on sandy rockless plains with shelter in rodent burrows, in varied situations in and under dead branches and trunks of Joshua trees, mesquites, sahuaros, oaks, etc., as well as under surface trash and litter almost anywhere in its range.

80. *Tantilla nigriceps* Kennicott. Plains Black-headed Snake. Southeastern part of the state, in the Sonoran Desert at its eastern edge and primarily in the desert-grassland. *Lower Sonoran* and *Upper Sonoran.* Secretive and rarely seen active on the surface, but readily found under rocks and stones in the desert-grassland and under surface litter, especially when the surface soil is damp in the spring (March-April) after winter rains, and in July-August after summer rains.

81. ***Tantilla atriceps*** Günther. Mexican Black-headed Snake. Southeastern part of the state, and in the northwestern part in the Grand Canyon; principally in the Sonoran Desert at its eastern edge and in the desert grassland. *Lower Sonoran* and *Upper Sonoran.* Locally abundant along rocky edges of washes, arroyos, and streams in desert valleys, and on rocky hillsides in the desert-grassland and lower oak woodland; secretive and rarely seen active on the surface, but readily found under rocks, rock rubble, dead agaves, yuccas, sotol, and other objects, especially when the surface soil is damp in spring (March-April) after winter rains, and in July-August after summer rains.

82. ***Tantilla wilcoxi*** Stejneger. Huachuca Black-headed Snake. Restricted to Sierra Madrean ranges ("Mexican Mountains") in the extreme southeastern part of the state; Patagonia and Huachuca Mountains, in woodland and grassland associations. *Upper Sonoran.* Found under rocks, logs, and dead plants such as yucca, agave, and sotol in shaded rocky canyons and on relatively open and sunny rocky slopes in the desert-grassland and evergreen woodland.

Family Elapidae: Cobras and Coral Snakes, *et al.*

83. ***Micruroides euryxanthus*** Kennicott. Arizona Coral Snake. Most of southern Arizona south of the Mogollon Rim, extending westward into Yuma County. In desert and grassland, and into lower woodland, to 5,800 feet. *Lower Sonoran* and *Upper Sonoran.* Fossorial, inoffensive, and little-studied in its natural environment due to its secretive habits and great amount of time spent underground; most abundant in the rocky Arizona Upland desert where there is a wide variety of closely adjacent soil types from loose sand to rock and rock rubble.

Family Crotalidae: Pit Vipers

84. ***Sistrurus catenatus*** Rafinesque. Massasauga. Rarely encountered in Arizona where it is known from a few scattered localities in grassland in the extreme southeastern corner of the state, in Cochise County. *Upper Sonoran.* Flat and undulating grass plains, often rockless; shelter in rodent burrows and in rock outcrops.

85. ***Crotalus atrox*** Baird and Girard. Western Diamondback Rattlesnake. Widely distributed in the desert and grassland, also into woodland in the southeastern and southwestern parts of the state, to approximately 5,300 feet. *Lower Sonoran* and *Upper Sonoran.* A wide variety of habitats; rock outcrops, boulder fields, rodent burrows, woodrat nests, plant thickets, and holes in arroyo banks are used for shelter and/or winter denning in the desert and desert-grassland; most abundant in the desert-grassland at elevations between 3,000 and 4,500 feet.

86. *Crotalus molossus* Baird and Girard. Black-tailed Rattlesnake. Widely distributed over most of the state on forested mountains and plateaus, and sparingly on lower desert ranges. Absent from the extreme northwest and northeast corners. *Lower Sonoran* to *Transition* (rarely *Canadian*) zones. Primarily rocky areas between 1,500 and 9,500 feet; this species has one of the widest elevational ranges exhibited by rattlesnakes in Arizona (see *C. viridis*), from desert into high coniferous forest.

87. *Crotalus scutulatus* Kennicott. Mohave Rattlesnake. Widely distributed in the deserts and grasslands of southern Arizona; very rarely as high as the lower edge of ponderosa pines in the westcentral part of the state, at 5,000-5,500 feet. *Lower Sonoran* and *Upper Sonoran* (not a woodland or forest species). Principally in open and non-rocky, flat or undulating terrain, and less frequently or absent on bajadas, fans, foothills, and hillsides where soils may be particularly rocky and vegetation relatively dense. Shelter primarily underground in rodent burrows, and above ground in rat nests, under boards, and trash.

88. *Crotalus mitchelli* Cope. Speckled Rattlesnake. Mohave and Sonoran Deserts in extreme southwestern and westcentral Arizona eastward to the Phoenix area in the Salt River basin. *Lower Sonoran.* Rock dweller primarily inhabiting canyons, buttes, and desert hills and ranges, rarely found beyond the limits of their rocky fans and bajadas.

89. *Crotalus tigris* Kennicott. Tiger Rattlesnake. Southcentral Arizona in the Sonoran Desert. *Lower Sonoran.* Strict rock dweller in rocky canyons and on hillsides and bajadas of desert ranges.

90. *Crotalus viridis* Rafinesque. Western Rattlesnake. Widely distributed in the state and conspicuously absent from the southwestern desert part (absent from almost all Mohave and Sonoran desert habitats). Five subspecies (races) in various colors and sizes occur from the Great Basin Desert in the north and grassland in the east into evergreen woodland and ponderosa pine forest mountain islands from the Santa Catalina and Rincon Mountains in the southeast (Pima County) to the Hualapai Mountains in the northwest (Mohave County). *Lower Sonoran* to *Transition.* Highly variable habitats between approximately 3,000 and 10,500 feet elevation, from the bottom of the Grand Canyon (rocks, crevices, beach drift) through grassland and evergreen woodland (rodent burrows, rock outcrops, wood rat nests) into coniferous forest (logs, rocks, etc.). This species extends over one of the widest elevational ranges exhibited by rattlesnakes in Arizona (see *C. molossus*).

91. *Crotalus lepidus* Kennicott. Rock Rattlesnake. Sierra Madrean ranges in the southeastern corner of the state: Santa Rita, Dragoon, Huachuca, and Chiricahua mountains. *Upper Sonoran.* Mid-elevation rock dweller primarily in rocky areas in the evergreen woodland (encinal and Mexican pine-oak) to about 6,300 feet; often common in rock-slides on south-facing slopes in the woodland above 5,000 feet. Occurs, in rocky areas, from the upper edge of the desert-grassland to the lower edge of the ponderosa pine forest (*Transition*).

92. *Crotalus pricei* Van Denburgh. Twin-spotted Rattlesnake. Sierra Madrean ranges in the southeastern corner of the state: Santa Rita, Huachuca, Dos Cabezas, Chiricahua, and Pinaleno (Graham) mountains. *Hudsonian* (spruce-fir forest), *Canadian* (white fir-Douglas fir forest), *Transition* (ponderosa pine forest), into the upper part of the *Upper Sonoran* (Mexican pine-oak woodland). Primarily in or near rocky areas in coniferous forest such as rock outcrops, rocky prominences of high peaks, and boulder rock-slides on south-facing slopes; from over 10,000 feet to as low as 6,300. Diurnal, crepuscular and nocturnal apparently according to season.

93. *Crotalus willardi* Meek. Ridge-nosed Rattlesnake. Santa Rita and Huachuca Mountains in the southeastern corner of the state. *Transition* downward to the *Upper Sonoran* above 5,600 feet. More commonly in or near crevices of rock outcrops on forest and woodland floors, as well as crawling on the often needle-covered ground principally under the canopy of pine trees; and in canyon bottoms with canopies of alder, box elder, maple, oak and other broadleaf deciduous trees. Diurnal, crepuscular, and nocturnal apparently according to season.

94. *Crotalus cerastes* Hallowell. Sidewinder; Horned Rattlesnake. Sonoran Desert and Mohave Desert in the southern and western parts of the state, west of an approximate line Phoenix-Florence-Marana; in the extreme south, the Tucson Mountains mark the eastward limit. *Lower Sonoran.* While usually found on sandy soils with rodent burrows serving as primary shelter, this strictly desert species makes a successful living in a number of different situations in Arizona below 2,300 feet elevation. It occurs on highly varied desert surfaces from barren fine-sand dunes, bare hardpan and rocky-gravelly desert "pavement," across scantily vegetated washes, sandy creosotebush flats, and playas and hummocks with densely clumped mesquite-dominated scrub, to rocky outwash slopes of paloverde-sahuaro foothills in the Arizona Upland desert.

PART 4

SPECIES OF BIRDS IN ARIZONA

Gale Monson
Bureau of Sport Fisheries and Wildlife
U. S. Fish and Wildlife Service,
Washington, D.C.

and

Allan R. Phillips
Instituto de Biología
Universidad Nacional Autónoma de México
México, D.F., México

Nearly half a century has elapsed since the appearance of the latest, and until now the only twentieth century, check list of the birds of Arizona. This was Harry S. Swarth's "A Distributional List of the Birds of Arizona," printed in 1914 by the Cooper Ornithological Club as No. 10 of the *Pacific Coast Avifauna* series. Twenty years later, in 1934, Anders H. Anderson brought the list up to date, as far as records in the literature were concerned, with "The Arizona State List Since 1914," published in *The Condor* (vol. 36, p. 78-83). Since these publications, a tremendous amount of work has been done both in the field and in museums, with the result that many concepts of Arizona's bird life have been radically changed, and the status of most species is now much more fully known.

The present list summarizes the known status of the Recent birds of Arizona up to and including 1960.* All published data have been incorporated, as well as considerable unpublished information accumulated by both authors, each of whom has spent more than twenty years in the state, much of the time in the field study of Arizona ornithology. Phillips has also checked most of the museums in the United States known to contain Arizona specimens, and most or all of the collections, public and private, in Arizona.

A knowledge of the vegetation, topography, and climate of Arizona is of course necessary to any understanding of the state's birds. There are a number of good texts on these subjects. Nevertheless, it is considered appropriate to present a resumé here, based on the principal life zones that are mentioned repeatedly in the following pages:

1. The *Lower Sonoran Zone*. This is the hot desert zone, whose typical plants are mesquite, palo verde, large cacti, and creosote bush. It covers much of southern and western Arizona, and follows the Colorado River up

*The authors have added a few significant records obtained subsequent to the end of 1960. [Editor]

to the bottom of the Grand Canyon. Elevation above sea level, 150 to 3,500 feet or more.

2. The *Upper Sonoran Zone.* This is also an essentially desert zone, but is not as arid and hot as the Lower Sonoran. Its typical plants are sagebrush and juniper-pinyon ("pygmy conifer") in the north, chaparral (brush) in the central part of the state, and live oaks of several species in the mountains of the southeast. Elevation, approximately 4,000 to 6,500 feet.

3. The *Transition Zone.* This is mainly a ponderosa pine forest, covering most of the Kaibab and Mogollon Plateaus, and large portions of the higher mountains elsewhere. Associated trees are Gambel's oak and New Mexico locust. Elevation, approximately 6,500 to 8,500 feet.

4. The *Canadian and Hudsonian Zones.* The Canadian Zone occupies only the higher elevations in the state, on the Kaibab Plateau and on the San Francisco, White, Graham, Chiricahua, and Chuska mountains plus high canyon heads in other mountain ranges. Its typical plants are the Douglas and white firs, limber pine, and quaking aspen. Elevation, approximately 8,500 to 10,000 feet. There are also small tracts above 9,500 feet that represent the Hudsonian Zone, but this zone is faunally much like the Canadian. Its typical plants are Engelmann spruce, cork-bark fir, and fox-tail pine. Because of the faunal similarity, both Canadian and Hudsonian Zones are commonly referred to herein as *boreal zones* or *boreal forests.*

On the San Francisco and White Mountains occur timberlines and touches of the *Arctic-Alpine Zone.*

The life zones are often intricately mixed, with intrusions of one zone into another, especially to canyons and broken mesa lands. Development of extensive tracts of irrigated farmland and especially the creation of large water impoundments on the major rivers have added variety to a bird habitat already amazingly diverse for a wholly inland, arid region. Elsewhere in Arizona, however, streams and marshy areas have largely disappeared, along with much of the native grassland.

Of particular interest to ornithologists are the many largely Mexican species of birds found in the southeastern part of Arizona, some of which (such as the Rose-throated Becard, Thick-billed Kingbird, Five-striped Sparrow) have apparently reached the state in the last decade or two.

The nomenclature and order of the 1957 (fifth) edition of the American Ornithologists' Union Check-list are used herein, although we do not agree with it in all respects. A minor departure is in the case of the family Tyrannidae, where we follow what is to us a more natural order of species within the genera, partly taken from Ridgway (*Birds of North and Middle America*), rather than the A.O.U. Check-list order. We have also indicated our disagreement with the A.O.U. Check-list in a few other instances (*i.e.*, flickers and juncos).

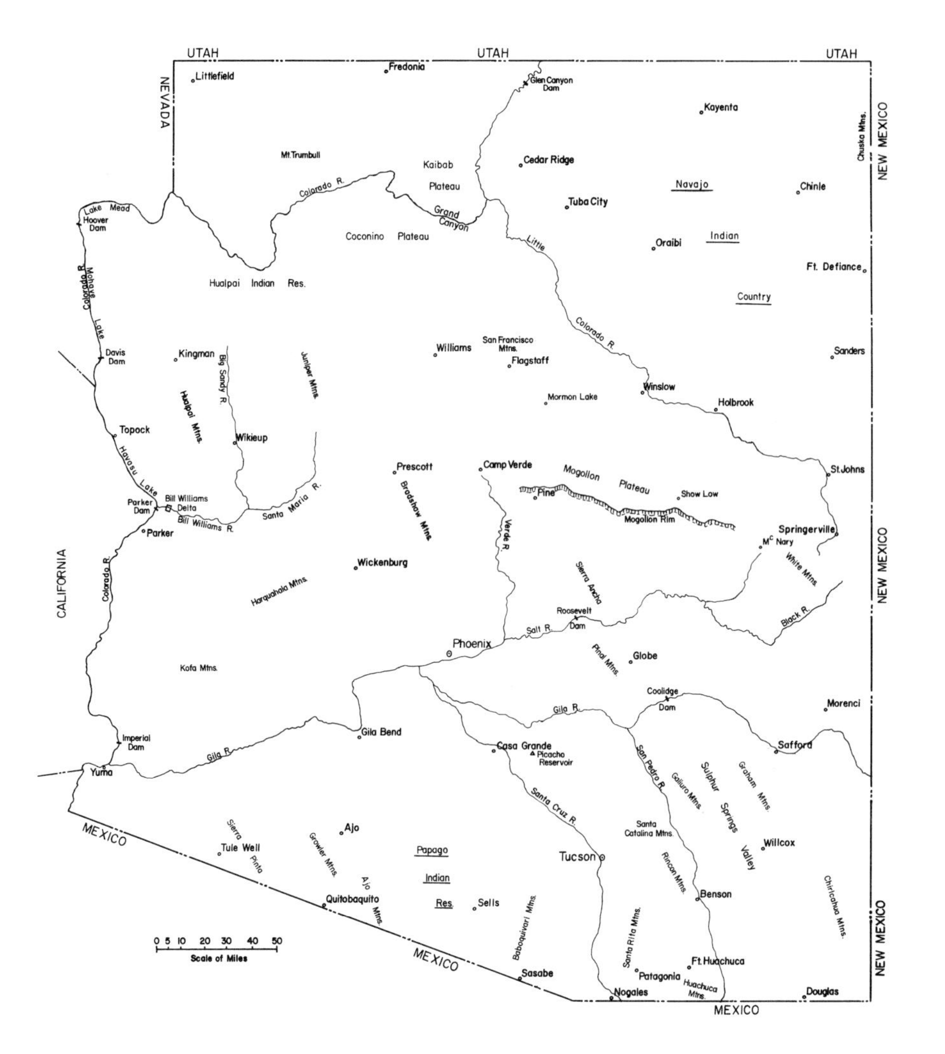

Map of ARIZONA Showing Principal Features and Cities.

Figure 1

The practice of admitting to the state list only those species represented by at least one collected specimen is followed. Where there is good reason to believe that other species have occurred in the state, on the basis of sight records by experienced ornithologists, these are included as hypothetical additions to the state list and are set apart from admitted species by being unnumbered and enclosed in brackets. We also cite records from adjacent parts of surrounding states, including Sonora, which we feel will serve to round out and complete a presentation of Arizona ornithology. The *dates* of records *substantiated by specimens are italicized.* The list contains 434 numbered full species as they are recognized by the A.O.U. Check-list (although not necessarily by us), which compares to Swarth's 1914 list of 332 currently-recognized full species.

As a rule, we have not attempted to more than indicate the seasonal status of each species. It should be borne in mind that the fall migration of many birds, especially water birds and hummingbirds, begins early in July, or even late in June, and that many mountain birds begin appearing in the lowlands at about the same time. Where the movements of any species are especially irregular, this is mentioned.

The reader will please bear in mind that since this is a list of *species,* and therefore is not concerned with *subspecies* (geographical races), all statements pertain to the species as a whole. Thus the entire statement regarding any particular species if often a composite of the status of two or more subspecies. Anyone reading it as an account of one individual bird may draw false conclusions. For example, birds migrating through the deserts of southwestern Arizona, or the mountain birds that occasionally visit the valleys of the southeast in winter, are frequently *not* of the same races that nest in our mountains, but come from other parts of North America. The easy assumption of "altitudinal migration" in such cases is clearly incorrect.

While the amount of work done on Arizona birds has been truly prodigious, much study remains to be accomplished, and farthest from our intentions is that the present list should be considered anything but a yardstick or foundation for future research. We would particularly like to point out the almost complete lack of sustained field work on the Kaibab Plateau and westward, on and about Lake Mead, and in the extreme northeast.

In stating the relative abundance of a given bird species it is difficult to find terms that will be uniformly interpreted. In this check-list, we have employed the following criteria: abundant — in numbers; common — always to be seen, but not in large numbers; fairly common — very small numbers, or not always seen; uncommon — seldom seen, but not a surprise; rare — always a surprise, but not out of normal range; casual — out of normal range; and accidental — far from normal range, and not to be expected again.

Any references to "A.O.U. Check-list" in the following pages are to the 1957 edition. The "Mormon Lake" referred to is that one about 20 miles southeast of Flagstaff, and is not to be confused with another lake of the same name near Show Low. The designation "lower Colorado River" or "lower Colorado Valley" applies to that portion of the stream below Davis Dam in the case of the land birds, and to that portion below the head of Lake Mead in the case of water birds.

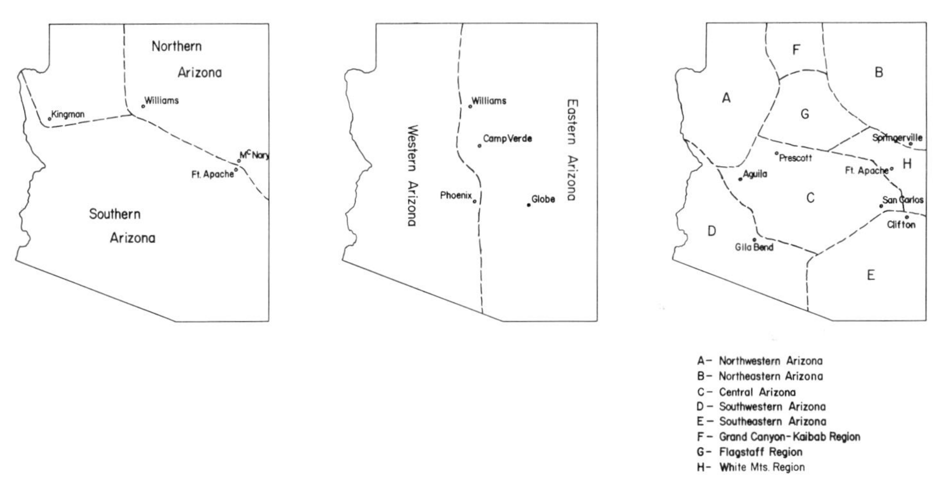

Figures 2, 3 and 4

Figure 1 (map) is for convenient reference for locating the principal physiographic features and ornithological localities. Figures 2, 3, and 4 present our conceptions of the various (northern, southeastern, etc.) sections of the state. Please bear in mind that any lines shown on these three maps are, to a considerable degree, necessarily arbitrary.

We wish to express our sincere thanks to the many persons who have been of assistance in the compilation of this check list. While their names are far too numerous to be set down here in full, we must especially mention the work of Robert W. Dickerman, Lyndon L. Hargrave, E. C. Jacot, Seymour H. Levy, and Joe T. Marshall, Jr. Ralph S. Palmer has been of help in bibliographical research. We are further grateful to the University of Arizona Press, and to Charles H. Lowe, whose efforts have made publication of the check list possible.

Gale Monson
Allan R. Phillips
December, 1963

Family Gaviidae: Loons

1. *Gavia immer* (Brünnich). Common Loon. Rare transient throughout Arizona, except on Colorado River lakes, where it is sometimes common in April and has been found every month of the year, including one at Havasu Lake, *July 28, 1954.*

2. *Gavia arctica* (Linnaeus). Arctic Loon; Pacific Loon. Very rare fall and early winter visitant; five specimens, north and east to Camp Verde and Tucson. Has been seen near Yuma and on Havasu Lake in spring, and on Havasu Lake on Aug. 13, 1954. Also recorded in southwestern Utah.

[***Gavia stellata*** (Pontoppidan). Red-throated Loon. Hypothetical. One near Ft. Mohave, November, 1947, and one on Havasu Lake June 8 and 15, 1948. We have been unable to verify a supposed specimen taken many years ago, and sight records for the Papago Indian Reservation and from near Tuba City are unsubstantiated.]

Family Podicipedidae: Grebes

3. *Podiceps auritus* (Linnaeus). Horned Grebe. Rare transient along Colorado River; several sight records but only one specimen, lower Havasu Lake, *Oct. 27, 1952.* Not yet certainly identified elsewhere in state; at least one alleged specimen proves to be *P. caspicus,* and others have been lost.

4. *Podiceps caspicus* (Hablizl). Eared Grebe. Fairy common to common transient statewide. Winters in small numbers in the lower Colorado and Salt River valleys and occasionally elsewhere, and a few may sometimes be found in summer. Nests on lakes in San Francisco and White Mountains regions, and possibly in Verde Valley.

5. *Podiceps dominicus* (Linnaeus). Least Grebe; Mexican Grebe. A straggler in recent years along the south edge of Arizona; recorded from Quitovaquito, Organ Pipe Cactus National Monument, *Apr. 28, 1939;* near Tucson, Dec. 28, 1941 and *Sept. 26, 1943;* and 10 miles north of Sasabe, *Jan. 14, 1958.* There is an old sight record for Camp Verde. Nested at California end of Imperial Dam (West Pond) in 1946.

6. *Aechmophorus occidentalis* (Lawrence). Western Grebe. More or less common transient and winter resident, rare to uncommon in summer, on lakes of the Colorado River; very unusual migrant elsewhere, unrecorded on the high plateaus.

7. *Podilymbus podiceps* (Linnaeus.) Pied-billed Grebe. Nests in marshy lakes scatteringly throughout state, casually even in late winter in the southeast, but commonly only in lower Colorado Valley and in high mountain country; winters in most ice-free waters.

Family Phaëthontidae: Tropic-birds

8. *Phaëthon aethereus* Linnaeus. Red-billed Tropic-bird. Accidental straggler; exhausted specimens have been found near Phoenix and in Apache Pass, near Dos Cabezas, Cochise County, *Apr. 10, 1905* and *Sept. 15, 1927*, respectively.

Family Pelecanidae: Pelicans

9. *Pelecanus erythrorhynchos* Gmelin. White Pelican. Regular transient, also wintering and summering in small numbers, along the lower Colorado River. Elsewhere occurs irregularly, subject to great fluctuations over the years but recently chiefly in small numbers or as singles. No verified breeding record.

10. *Pelecanus occidentalis* Linnaeus. Brown Pelican. A few enter the state in summer and fall, chiefly in recent years, rarely staying until winter or even spring; most records are for the Colorado Valley, but stragglers reach most of state, exceptionally even the northeast.

Family Sulidae: Boobies and Gannets

11. *Sula nebouxii* Milne-Edwards. Blue-footed Booby. Rare late summer and fall straggler to lower Colorado Valley, at least in 1953 and 1954, north to Havasu Lake; also found once at Phoenix, *July 29, 1953.*

12. *Sula leucogaster* (Boddaert). Brown Booby; Brewster's Booby. Rare late summer and fall straggler to lower Colorado Valley, north to Havasu Lake. One present continuously from early September, 1958 to October, 1960 at Martinez Lake about seven miles above Imperial Dam.

Family Phalacrocoracidae: Cormorants

13. *Phalacrocorax auritus* (Lesson). Double-crested Cormorant; Farallon Cormorant. Common transient and winter resident in lower Colorado Valley, where it breeds locally. Occurs in smaller numbers on Lakes Mohave and Mead and on larger lakes of the Salt River, and rarely elsewhere in the state, where chiefly a transient.

13a. *Phalacrocorax olivaceus* (Humboldt). Olivaceous Cormorant; Mexican Cormorant. A pair found at Arivaca Junction, Pima County, *Mar. 10, 1961.*

Family Anhingidae: Anhingas

14. *Anhinga anhinga* (Linnaeus). Anhinga; Water-Turkey. Accidental; only one certain record, from Tucson, *Sept. 12, 1893.* Old sight records at Yuma, where it perhaps occurred about 1900, and at California end of Laguna Dam, Feb. 9, 1913, and more recently, but none substantiated.

Family Fregatidae: Frigate-birds or Man-o'-War Birds

[***Fregata sp.*** Frigate-bird (species?). Hypothetical. Seen twice at Havasu Lake, Aug. 13, 1954, and Sept. 8, 1955.]

Family Ardeidae: Herons and Bitterns

15. *Ardea herodias* Linnaeus. Great Blue Heron; Treganza's Heron. Found at any season, but in small numbers, throughout the state wherever there is much open water, nesting in scattered small heronries.

16. *Butorides virescens* (Linnaeus). Green Heron. Breeds uncommonly along streams, particularly in willow areas, south and west of the Mogollon Plateau; found sparingly in winter in same areas, except northernmost parts, chiefly in recent years (since 1940).

17. *Florida caerulea* (Linnaeus). Little Blue Heron. Casual; one wounded at Camp Verde, July 27, 1885, and one taken near Phoenix, *May 27, 1958.* A sight record for Lake Mead in Nevada is dubious.

18. *Dichromanassa rufescens* (Gmelin). Reddish Egret. Casual; one specimen from near Camp Verde, *Aug. 27, 1886.* Also, one seen repeatedly along Colorado River above Imperial Dam, September 1954 to March 1955; one on California side of Havasu Lake, *Sept. 9, 1954;* and one about 30 river miles above Imperial Dam, Sept. 2, 1960.

19. *Casmerodius albus* (Linnaeus). Common Egret; American Egret. Common resident, breeding in scattered colonies, along lower Colorado River; fairly common winter resident at Picacho Reservoir. Elsewhere found in very small transient numbers, a few wintering south of Mogollon Plateau. Reports of its nesting on or north of Mogollon Plateau, or in central Arizona, are not based on valid evidence. May be found in central Arizona all summer, except perhaps end of June and first three weeks of July.

20. *Leucophoyx thula* (Molina). Snowy Egret; Brewster's Egret. Found year-long in Colorado Valley, especially common as transient and in recent years as a winter resident; has nested in the Topock Swamp and on California side of river at Taylor Lake, 17 miles above Imperial Dam. Elsewhere a frequent migrant, in most places in small flocks or singly. Occasional in central and southeastern Arizona all summer, except perhaps end of June and early July. Only one winter record east of lower Gila Valley: Benson, Jan. 31, 1940.

21. *Hydranassa tricolor* (Müller). Louisiana Heron; Tri-colored Heron. Very rare fall straggler in southwestern and central Arizona; has been found north to Camp Verde, *Sept. 24, 1884.* A winter report from Lake Mead, Nevada, would seem to require specimen support.

22. *Nycticorax nycticorax* (Linnaeus). Black-crowned Night Heron. Resident in Colorado Valley, where it breeds; sometimes seen elsewhere as transient, occasionally in waterless situations, or as winter visitant in southern and central Arizona. Bred formerly on Salt and Verde Rivers, but present status there uncertain; may breed near Springerville.

23. *Nyctanassa violacea* (Linnaeus). Yellow-crowned Night Heron. Accidental: one found at Sullivan Lake, north of Prescott, *Oct. 14, 1951.*

24. *Ixobrychus exilis* (Gmelin). Least Bittern. Uncommon but found year-long, principally in summer, in lower Colorado Valley; rare in summer in marshy places of central Arizona, where it has bred. Elsewhere rare transient or wanderer and no winter records, except at Picacho Reservoir, Pinal County, in 1942-43.

25. *Botaurus lentiginosus* (Rackett). American Bittern. Uncommon from late August to May in cattail areas in lower Colorado Valley; rare elsewhere and chiefly a transient. Before 1915 nested at lakes on Mogollon Plateau; no winter records for that region or northward.

Family Ciconiidae: Storks and Wood Ibises

26. *Mycteria americana* (Linnaeus). Wood Ibis. Frequent summer visitant (non-breeding) along lower Colorado and Gila Rivers from mid-June to early October; uncommon elsewhere in southern part of state, and rare on and north of Mogollon Plateau. No winter records since 1903, and no spring record since 1910.

Family Threskiornithidae: Ibises and Spoonbills

27. *Plegadis chihi* (Vieillot). White-faced Ibis; Glossy Ibis; White-faced Glossy Ibis. Fairly common migrant, perhaps less common in northeast; a few (rarely up to about 50) may be found throughout the summer, but no positive breeding record. Rare in winter in southern Arizona.

[***Eudocimus albus*** (Linnaeus). White Ibis. Hypothetical. An adult seen at Martinez Lake, Yuma County, April 4, 1962, and satisfactorily photographed the next day. One reported seen on California side of Colorado River at Palo Verde, Imperial County, March 1914, but no substantiation.]

[***Eudocimus ruber*** (Linnaeus). Scarlet Ibis. Hypothetical. While there is no specimen, the identification of a flock of seven or eight near Tucson, Sept. 17, 1890, by so experienced a collector as Herbert Brown can hardly be questioned in the case of so striking a bird.]

28. *Ajaia ajaja* (Linnaeus). Roseate Spoonbill. Casual in southern Arizona from Yuma east to near Phoenix; records extend from May to November, including one at Quitovaquito, Organ Pipe Cactus National Monument, May 30, 1951.

Family Anatidae: Swans, Geese, and Ducks

29. *Olor columbianus* (Ord). Whistling Swan. Winters uncommonly in lower Colorado Valley (and in Big Sandy Valley?); now rare elsewhere.

30. *Branta canadensis* (Linnaeus). Canada Goose; "Hutchins' Goose." Winters more or less commonly in southwestern and central Arizona and along the Virgin River; uncommon migrant in northern Arizona. The race *minima* Ridgway, Cackling Goose, should perhaps be assigned to a separate species, *B. hutchinsii* (Richardson); it has been taken casually at Topock, Colorado River, *Dec. 15, 1949,* and *Nov. 14, 1953.*

[***Branta sp.*** Brant (species?). Hypothetical. One examined by Jacot and Vorhies at Picacho Reservoir, Pinal County, about November, 1941 or 1942. It was muddy and bedraggled and was not saved. Old reports, and recent ones from Lake Mead, are unlikely and unsubstantiated.]

31. *Anser albifrons* (Scopoli). White-fronted Goose. Now rare to uncommon; formerly a common fall transient in the Colorado Valley and in smaller numbers elsewhere. Winter and even spring records are not numerous and are chiefly from central-southern Arizona.

32. *Chen "hyperborea* (Pallas)." Snow Goose. Formerly common winter visitant to Colorado and Gila valleys, now reduced in numbers; still winters on lower Colorado River, and in small numbers in Salt River Valley east to Roosevelt Lake. Elsewhere an uncommon, almost rare, and local fall migrant, with only two winter records (bottom of Grand Canyon, Jan. 3 to 9, 1937, and St. Johns, one shot Jan. 11, 1951).

[***Chen caerulescens*** (Linnaeus). "Blue Goose." Probably a color phase of the Snow Goose. If considered distinct, hypothetical: one seen at Topock, Jan. 20-31, 1950, and another Dec. 29, 1953 and during Jan. 1954.]

33. *Chen rossii* (Cassin). Ross' Goose. Casual. Three records, one near Camp Verde, *Oct. 24, 1887,* one near Topock, *Dec. 10, 1948,* and four adults at Martinez Lake above Imperial Dam, Nov. 5 to at least Dec. 31, 1960.

34. *Dendrocygna autumnalis* (Linnaeus). Black-bellied Tree Duck. Of sporadic and rare occurrence in summer in southeastern and central Arizona; has nested near Tucson, and at Hereford, San Pedro Valley. Two families were wintering near Nogales in December, 1960.

35. *Dendrocygna bicolor* (Vieillot). Fulvous Tree Duck. Now rare from spring to early winter in southern and western Arizona; recent records are almost entirely of singles or pairs. Formerly a fairly common winter resident at Yuma. No breeding record.

36. *Anas platyrhynchos* Linnaeus. Mallard. Common transient and winter resident wherever there is open water. Breeds in high mountain lakes in the north, and locally in the lower Colorado Valley and (formerly) near Phoenix.

37. *Anas "diazi"* Ridgway. "Mexican Duck," "New Mexico Duck." Found locally in summer, at least, in extreme southeast; definitely known to nest at San Bernardino Ranch east of Douglas, and in San Simon Cienega (at least in New Mexico part). Straggler west to Salt River (one specimen, *about 1943*). (Probably this and the "Mottled" and "Black Ducks" are races of the Mallard.)

38. *Anas strepera* Linnaeus. Gadwall. Fairly common transient, and winters along low river valleys. Nests on high mountain lakes, and locally and irregularly in very small numbers in the lower Colorado Valley. Non-breeding individuals occasional elsewhere in summer.

39. *Anas acuta* Linnaeus. Pintail; Sprig. Abundant to common fall migrant, winters on open waters; rather uncommon spring migrant. Nests on lakes of the San Francisco Mountains region. Non-breeding birds may be seen elsewhere in summer.

[***Anas crecca*** Linnaeus. Common Teal. Hypothetical. A male seen at Picacho Reservoir, Pinal County, Jan. 18, 1953. Possibly conspecific with *A. carolinensis.*]

40. *Anas carolinensis* Gmelin. Green-winged Teal. Abundant to common transient, and winter resident on open waters. Nests rarely on lakes of Mogollon Plateau, and rumored to have bred on Gila River.

41. *Anas discors* Linnaeus. Blue-winged Teal. Spring transient throughout state, numbers and dates variable from year to year. Fall status uncertain, but recorded to *Oct. 31* in northern Arizona and Nov. 17 along lower Colorado River; teals with blue wings are common fall migrants statewide. Sometimes seen in summer, even in south, and quite possibly nests on Mogollon Plateau.

42. *Anas cyanoptera* Vieillot. Cinnamon Teal. Common migrant in spring, and probably in early fall, although status uncertain then. Small numbers may occasionally winter in extreme southwest. Breeds commonly in lakes of the Mogollon Plateau, and rarely in southern Arizona.

[***"Mareca" penelope*** (Linnaeus). European Widgeon. Hypothetical. A male seen at Topock, Dec. 18, 1947.]

43. *"Mareca" americana* (Gmelin). American Widgeon; Baldpate. Fairly common migrant, wintering on open waters; non-breeding birds may be seen in summer. Prior to 1929 nested at least occasionally on lakes of the Mogollon Plateau.

44. *"Spatula" clypeata* (Linnaeus). Shoveler; Spoonbill. Common transient, wintering on open waters. Non-breeding birds or transients may be seen during summer. No positive nesting record.

45. *Aix sponsa* (Linnaeus). Wood Duck. Very rare, found almost throughout state, mostly in fall and early winter; one summer record, a male near Topock, June 7, 1946. Southernmost records are at Tucson and Arivaca.

46. *Aythya americana* (Eyton). Redhead. Fairly common transient. An occasional bird winters in southern Arizona, especially in lower Colorado Valley. Nests locally in the Colorado Valley and on the Mogollon Plateau. Non-breeding birds may be seen during summer.

47. *Aythya collaris* (Donovan). Ring-necked Duck. Rather uncommon migrant, sometimes common in fall; locally common in winter on open waters. Fairly common as a nesting bird locally in the White Mountains. Non-breeding birds may be seen rarely elsewhere in summer.

48. *Aythya valisineria* (Wilson). Canvasback. Rather uncommon migrant, and winters in some numbers locally on open waters; sometimes numerous in fall on mountain lakes. There are a few summer records.

49. *Aythya marila* (Linnaeus). Greater Scaup. Probably more common and widespread than the only two verified records would indicate: one from near Clarkdale, *Dec. 8, 1887*, and a partial skeleton found at Picacho Reservoir, Pinal County, *Jan. 31, 1953*. Also one reported from lower Colorado River (in Baja California), *Feb. 7, 1928*.

50. *Aythya affinis* (Eyton). Lesser Scaup. Common to abundant migrant, especially in fall; winters occasionally wherever there is open water, even on the Mogollon Plateau lakes. Occasional non-breeding birds may be seen throughout the summer.

51. *Bucephala clangula* (Linnaeus). Common Goldeneye; American Goldeneye. Rare to uncommon winter visitant on unfrozen lakes, chiefly of central and western Arizona; sometimes locally numerous along Colorado River. Fall migrant on Mogollon Plateau and north. Individuals or pairs may remain until May.

[***Bucephala islandica*** (Gmelin). Barrow's Goldeneye. Hypothetical. An old specimen supposedly taken at Phoenix. We have not located it for re-examination.]

52. *Bucephala albeola* (Linnaeus). Bufflehead. Still a frequent transient; in winter fairly common in Colorado Valley but much scarcer eastward, especially in recent years. Casual near Phoenix, June 20, 1943.

53. *Clangula hyemalis* (Linnaeus). Oldsquaw. Rare fall and winter visitant, recorded south to Tucson, Arlington, and near Yuma.

54. *Melanitta perspicillata* (Linnaeus). Surf Scoter. Casual; two specimens — one from Hillside, west of Prescott, *Oct. 20, 1929,* and one from Havasu Lake, *Oct. 23, 1949.* Also fall sight records, one each from Topock and Havasu Lake and two (one photographed) from near Yuma, and one in March from Canoa Ranch, southern Pima County.

55. *Oxyura jamaicensis* (Gmelin). Ruddy Duck. Common transient in lower Colorado Valley, less common in eastern Arizona. Winters in small numbers almost wherever there is open water. Breeds on Mogollon Plateau, locally along Colorado River, rarely elsewhere.

56. *Lophodytes cucullatus* (Linnaeus). Hooded Merganser. Winter resident in very small numbers, principally in lower Colorado Valley above Parker Dam. One summer record: Santa Cruz River near Calabasas, *June 1, 1890* (if correctly re-labelled — possibly *January 6?*).

57. *Mergus merganser* Linnaeus. Common Merganser; American Merganser. Common winter resident on larger bodies of open water, uncommon elsewhere. Uncommon fall transient on lakes of Flagstaff area. Breeds in small numbers along streams below Mogollon Plateau and White Mountains. Occasional individuals may remain summer-long in the Colorado Valley.

58. *Mergus serrator* Linnaeus. Red-breasted Merganser. Irregular transient in Colorado Valley, sometimes common; elsewhere decidedly uncommon in recent years. Occasional individuals may remain winter or summer-long in the Colorado Valley. Formerly wintered abundantly on Roosevelt Lake, and a few wintered at Lakeside, Navajo County, on the Mogollon Plateau.

Family Cathartidae: American Vultures

59. *Cathartes aura* (Linnaeus). Turkey Vulture. Common summer resident, except in extensive forested areas where it is rather uncommon or locally rare. Winters in agricultural areas in Colorado Valley north to at least Topock, less commonly in Gila Bend and Phoenix areas and the Altar Valley, and casually east to Tucson.

60. *Coragyps atratus* (Bechstein). Black Vulture. Resident in recent years in small numbers of Santa Cruz Valley above Picacho Peak, and probably the Papago Indian Reservation west (irregularly?) to Organ Pipe Cactus National Monument. Occasionally seen north to Arlington and Gila River Indian Reservation area; recorded casually from Wickenburg, north of Oracle Junction in Pinal County, and extreme southeastern Arizona. Supposed records for Tonto Basin, Verde Valley, and north are not considered authentic.

61. ***Gymnogyps californianus*** (Shaw). California Condor. A few sight records, principally in the 1880's, from southeast to northwest (and southwestern Utah). Most of the few bird bones (not petrified) recently recovered from caves in the Grand Canyon are of this species.

Family Accipitridae: Hawks, Old World Vultures, and Harriers

62. ***Accipiter gentilis*** (Linnaeus). Goshawk; American Goshawk. Rare to uncommon resident of the mountains of eastern and central Arizona and the Kaibab Plateau. In winter it occurs rarely well away from the pine forests, and irregularly so on the desert, even reaching the Colorado Valley in major flights (Palo Verde, California, *1916*).

63. ***Accipiter striatus*** Vieillot. Sharp-shinned Hawk. Uncommon summer resident of the mountains of eastern Arizona. Winters commonly in southern and western Arizona, uncommonly on Mogollon Plateau and farther north where it is mainly a transient.

64. ***Accipiter cooperii*** (Bonaparte). Cooper's Hawk. Nests scatteringly throughout the state. Winters commonly in southern and western Arizona except in higher parts of mountains. Transient in most or all parts of state.

65. ***Buteo jamaicensis*** (Gmelin). Red-tailed Hawk. Common resident almost statewide. Winters to upper part of Transition Zone and possibly higher.

66. ***Buteo "harlani"*** (Audubon). "Harlan's" Hawk. Casual; two records, one 10 miles south of St. David, *Dec. 4, 1959*, and one northwest of Chandler, *Jan. 10, 1962*. Regarded as a race of the Red-tailed Hawk, or a localized color-phase.

67. ***Buteo lineatus*** (Gmelin). Red-shouldered Hawk; Red-bellied Hawk. One specimen; Little Colorado River near Holbrook, *Dec. 5, 1853* (figured in Baird *et al.*, Birds of North America, 1860, plate 3). A few subsequent sight records from northeastern Arizona.

68. ***Buteo platypterus*** (Vieillot). Broad-winged Hawk. Casual; one specimen, Southwestern Research Station, Chiricahua Mountains, *Sept. 22, 1956*.

69. ***Buteo swainsoni*** Bonaparte. Swainson's Hawk. Common summer resident of grassy plains of eastern Arizona; also, but sparingly, to central and central-southern Arizona, and possibly to eastern Mohave County. A migrant state-wide except in forests, but rare in western Arizona. Casual in winter, most reports dubious and no specimens between October and March. An old breeding record near Yuma is not considered authentic, nor do we believe it a "common migrant," starting in February, in adjacent parts of Nevada.

70. *Buteo albonotatus* Kaup. Zone-tailed Hawk. Fairly common summer resident of mountains in northwestern, central and southeastern Arizona; has also nested in the Santa Cruz and Rillito valleys, on the Gila River, and near mouth of Bill Williams River (1946 and 1947); formerly commoner near Tucson. May possibly nest irregularly west to Organ Pipe Cactus National Monument, where individuals seen May 9 and 13, 1939. Occurs rarely in migration west to Colorado River. A very few winter records for Tucson region and some old ones from Yuma.

71. *Buteo albicaudatus* Vieillot. White-tailed Hawk. Casual; one or two specimens from Phoenix, *1899;* nest recorded from between Florence and Red Rock, 1897; one seen repeatedly near Marinette, Maricopa County, winter of 1954-55, and three the same winter west of Gila Bend.

72. *Buteo lagopus* (Pontoppidan). Rough-legged Hawk; American Rough-legged Hawk. Rare in winter, but probably of statewide occurrence, except in southwest. Specimens south to Sulphur Springs Valley at Sonora border.

73. *Buteo regalis* (Gray). Ferruginous Hawk; Ferruginous Rough-leg. Uncommon, but widely distributed, summer resident of grassy plains of northern Arizona, and locally and irregularly in southeastern Arizona. Fairly common in winter in north and southeast, relatively rare elsewhere.

74. *Buteo nitidus* (Latham). Gray Hawk; Mexican Goshawk. Formerly a fairly common summer resident along wooded streams in southeast, now probably restricted to Santa Cruz River above Tucson (where irregular) and the Sonoita Creek-Nogales area. One old winter sight record. No authentic record as far north as the Gila River.

75. *Parabuteo unicinctus* (Temminck). Harris' Hawk. Locally resident along Colorado River north to Topock, eastward to Superior, east of Florence, and Tucson, and southward to Mexican boundary on Papago Indian Reservation and Organ Pipe Cactus National Monument. Rarely seen in migratory flocks, usually with Swainson's Hawks, in Tucson region and westward. Formerly occurred to upper San Pedro Valley, casually to Verde Valley (and also extreme southeastern Arizona?).

76. *Buteogallus anthracinus* (Deppe). Black Hawk; Mexican Black Hawk. Regular summer resident along permanent streams in southeastern and central Arizona as far northwest as the Big Sandy drainage; also one-time sight records from Cataract Canyon southwest of Grand Canyon and near Parker. Winter sight records are not considered authentic. There is no factual basis for the statement "possibly resident" (A.O.U. Check-list).

77. *Aquila chrysaëtos* (Linnaeus). Golden Eagle. Sparingly distributed throughout state in all mountain areas; somewhat more common in the lower mountains in winter, when it also occurs along the Colorado River. Virtually absent in summer, after breeding, in some desert regions (*viz.*, lower Little Colorado Valley), from early May to August or mid-September.

78. *Haliaeetus leucocephalus* (Linnaeus). Bald Eagle. Not uncommon resident about lakes and streams of the White Mountains region, rarer west down to Salt River and to Flagstaff region; said to leave after nesting, at least on Salt River. Transient in northern mountains and Kaibab Plateau. Winter records mainly from the Flagstaff and Colorado River regions. Rare in southeast.

79. *Circus cyaneus* (Linnaeus). Marsh Hawk. Common winter resident of well-vegetated open areas below Transition Zone, except in northeast where it is mainly a transient. Before 1890 nested in parts of eastern Arizona, and may still do so west of Holbrook, but no positive record. Summer reports along North Rim of Grand Canyon probably do not indicate breeding. One seen at Topock June 12 and July 25, 1950.

Family Pandionidae: Ospreys

80. *Pandion haliaëtus* (Linnaeus). Osprey; Fish Hawk. Nests locally along streams below Mogollon Rim, also one possible nesting record from Mohave Lake area (1950). Found rarely in summer on Mogollon Plateau, and along Colorado River and upper Salt and Verde Rivers. May occur almost anywhere in migration. Occasionally winters in lower Colorado and Gila valleys, and formerly did so at Roosevelt Lake.

Family Falconidae: Caracaras and Falcons

81. *Caracara cheriway* (Jacquin). Caracara; Audubon's Caracara. Resident in small numbers on the Papago Indian Reservation; casual east to Santa Cruz Valley, west to Gila Bend and extreme southeastern Yuma County. Old records extend to Santa Cruz and Salt River Valleys, Florence, Oracle, and Yuma. A recent record near Pearce, Sulphur Springs Valley, March 1954.

82. *Falco mexicanus* Schlegel. Prairie Falcon. Scarce resident statewide, also winter visitant to southern and western Arizona. Formerly commoner.

83. *Falco peregrinus* Tunstall. Peregrine Falcon; Duck Hawk. Nests rarely at cliffs throughout state, even at some distance from water. Found statewide in migration; winters occasionally along lower Colorado River and in central Arizona.

84. ***Falco femoralis*** Temminck. Aplomado Falcon. Before 1890 a fairly common summer (or permanent?) resident in the southeast; since then virtually extinct in state, with only two exact records since before 1910: near McNeal, Cochise County, Nov. 13, 1939, and St. David, Oct. 7, 1940.

85. ***Falco columbarius*** Linnaeus. Pigeon Hawk; Merlin. Uncommon to rare transient and winter resident virtually statewide. Summer reports unsubstantiated.

86. ***Falco sparverius*** Linnaeus. Sparrow Hawk. Resident and transient, abundant in some areas in migration; uncommon to rare in winter in the higher mountains, and in summer in western Arizona.

Family Tetraonidae: Grouse and Ptarmigan

87. ***Dendragapus obscurus*** (Say). Blue Grouse; Dusky Grouse. Fairly common resident of the White Mountains region, in boreal zones, less common along the North Rim of the Grand Canyon and in the Chuska Mountains of northern Apache County.

[***Centrocercus urophasianus*** (Bonaparte). Sage Grouse. Hypothetical. One seen near Nixon Spring in the Mt. Trumbull region, July 29, 1937.]

Family Phasianidae: Quails, Pheasants, and Peacocks

88. ***Colinus virginianus*** (Linnaeus). Bobwhite; Masked Bobwhite. Extinct in Arizona for many years; formerly (mainly before 1890) common in tall grass-mesquite plains from Baboquivari Mountains east to upper Santa Cruz Valley. Grazed out of existence by early 1900's; attempted reintroductions have all failed. Records for Huachuca and Whetstone Mountains are not well-founded, though repeated in A.O.U. Check-list.

89. ***Callipepla squamata*** (Vigors). Scaled Quail. Common resident of grassy plains in southeastern and southcentral Arizona, west and northwest to Baboquivari Mountains, southcentral Pinal County, and east of San Carlos. Occurs locally in upper Little Colorado drainage, perhaps by introduction. Hybrids with the following species have been found occasionally in recent years, chiefly in the vicinity of Antelope Peak southwest of Winkelman.

90. ***Lophortyx gambelii*** Gambel. Gambel's Quail; Desert Quail. Abundant resident in all areas where mesquite occurs, locally higher (along foot of Mogollon Plateau). Native occurrence in northern Arizona not substantiated. Also introduced in various places in the northern part of the state, for the most part unsuccessfully.

91. ***Cyrtonyx montezumae*** (Vigors). Harlequin Quail; Mearns' Quail; Fool Quail. Uncommon to fairly common resident of grassy open woods of the mountains of southeastern and central Arizona, west to the Baboquivari Mountains and north to the Pinal and White Mountains regions;

formerly ranged sparingly to Flagstaff and Prescott areas. Found principally in Upper Sonoran Zone, but has occurred casually to Hudsonian Zone in summer.

[***Phasianus colchicus*** Linnaeus. Ring-necked Pheasant. Has been introduced repeatedly in agricultural areas, without success; it persists only by continued artificial replenishment. The statement (A.O.U. Check-list) that it is established in southeastern Arizona is without any basis known to us.]

[***Alectoris sp.*** Chukar. Has been introduced in many parts of the state, without any degree of success unless possibly in the Springerville or Virgin Mountains areas.]

Family Meleagrididae: Turkeys

92. *Meleagris gallopavo* Linnaeus. Turkey. In early days, an abundant resident of virtually all forested parts of state except Hualapai (and Galiuro?) Mountains and Kaibab Plateau, descending to some of the valleys in winter. By 1930 shot out except in San Francisco and White Mountains regions. Since then restocked over most of former range and elsewhere with some success, especially in southeastern mountains. We regard as doubtful the statement that it formerly nested in a swamp at Calabasas.

Family Gruidae: Cranes

93. *Grus canadensis* (Linnaeus). Sandhill Crane; Little Brown Crane. Formerly locally common; now distinctly uncommon and irregular winter resident in irrigated tracts of central and southern Arizona. Prior to 1936, a few northern Arizona records, including summer occurrences. Virtually unknown as a migrant in recent years. "Abundant" below Yuma, Apr. 9, 1862.

Family Rallidae: Rails, Gallinules, and Coots

94. *Rallus longirostris* Boddaert. Clapper Rail; Yuma Rail; Light-footed Rail. Summer resident (irregular?) in some alkaline cattail marshes along lower Colorado River, north to Bill Williams Delta. No authentic winter records.

95. *Rallus limicola* Vieillot. Virginia Rail. Summer resident of marshes in the White Mountains region down to Upper Sonoran Zone, and probably breeds in certain Maricopa County marshes. Locally a common migrant, at least along Colorado River, and formerly at Tucson. Rare to uncommon during winter, when it has been found as high as McNary, Apache County.

96. *Porzana carolina* (Linnaeus). Sora; Carolina Rail. Breeds commonly at marshes in northern Arizona, also probably in central Arizona. Common transient virtually statewide. Winters commonly at unfrozen marshes; recorded twice from frozen marshes in southern Navajo County.

97. *Coturnicops noveboracensis* (Gmelin). Yellow Rail. One caught alive near Sacaton, Pinal County, *Mar. 28, 1909*, the only substantiated record, though not improbably a rare winter visitant.

[***Laterallus jamaicensis*** (Gmelin). Black Rail. Hypothetical. Sight records by competent collectors in central-southern Arizona; Tucson (Stephens, Apr. 23, 1881) and near Casa Grande (D. D. Stone), but they were unable to capture the birds.]

98. *Porphyrula martinica* (Linnaeus). Purple Gallinule. Casual summer visitant (July to September) to the Tucson area, where there are several records; also one record for western Santa Cruz County, Montana Lake, near Oro Blanco, *Aug. 2, 1909*, and one caught alive in Arizona near Rodeo, N. Mex., June 1935. Occurrence elsewhere doubtful.

99. *Gallinula chloropus* (Linnaeus). Common Gallinule; Florida Gallinule; Black Gallinule. Common to fairly common resident locally across the southern and western parts of the state where cattails or tules are associated with shallow streams, ponds, or canals; somewhat less common in winter. Occasional migrant in less favorable spots.

100. *Fulica americana* Gmelin. American Coot; Mud-hen. Common summer resident, and common to abundant migrant, of water areas with cattail-tule margins. Winters commonly wherever open water is present, abundantly along lower Colorado River.

Family Charadriidae: Plovers, Turnstones, and Surfbirds

101. *Charadrius semipalmatus* Bonaparte. Semipalmated Plover; Ringed Plover. Uncommon transient along lower Colorado River, elsewhere rather rare; hardly any records in north and east.

102. *Charadrius alexandrinus* Linnaeus. Snowy Plover. Fairly common transient along lower Colorado River. Two wintered in Bill Williams Delta, 1952-53, and two above Imperial Dam, December 1960. Eastward, only one record, near Peoria, *Sept. 10, 1957*.

103. *Charadrius vociferus* Linnaeus. Killdeer. The ubiquitous waterbird of the state, found year-long anywhere at open water. Abundant at times in migration, and in irrigated areas in winter.

104. *Eupoda montana* (Townsend). Mountain Plover. Mostly a rare bird, in fall in small numbers in the western part of the state, and as wintering flocks on barren desert flats and fallow fields in the Florence,

Phoenix, Yuma and Kingman regions; very rare farther east, where there are no recent records. Several flocks seen northeast of Springerville in August, 1914, suggesting possible breeding.

105. ***Pluvialis dominica*** (Müller). American Golden Plover. One record: near Peoria, Maricopa County, *May 15, 1953.*

106. ***Squatarola squatarola*** (Linnaeus). Black-bellied Plover. Rare transient probably statewide, with most records from the lower Colorado Valley and none from northeast as yet. One winter record, at Yuma, Dec. 30, 1940.

[***Arenaria interpres*** (Linnaeus). Ruddy Turnstone. There are two records for the California side of Havasu Lake: *Sept. 16, 1952* and Aug. 21, 1953.]

[***Arenaria melanocephala*** (Vigors). Black Turnstone. One seen on the California side of Havasu Lake, May 21, 1948.]

Family Scolopacidae: Woodcock, Snipes, and Sandpipers

107. ***Capella gallinago*** (Linnaeus). Common Snipe; Wilson's Snipe. Fairly common to common transient throughout state, wintering generally at unfrozen waters. Breeds near Springerville.

108. ***Numenius americanus*** Bechstein. Long-billed Curlew. Uncommon transient statewide, more common in lower Gila and Salt River Valleys and along lower Colorado River. Very rare in winter in south, and even in the bottom of the Grand Canyon; occasional along lower Colorado River in summer, when it has also been found in the north.

109. ***Numenius phaeopus*** (Linnaeus). Whimbrel; Hudsonian Curlew. Very rare along lower Colorado River; one specimen, Bill Williams Delta, *Sept. 17, 1946,* plus three sight records. No verifiable report elsewhere.

110. ***Bartramia longicauda*** (Bechstein). Upland Plover; Bartramian Sandpiper. A few transient records from Pima, Cochise, and Graham Counties, the last in *May, 1887.* One seen on California side of Havasu Lake, Sept. 11, 1952.

111. ***Actitis macularia*** (Linnaeus). Spotted Sandpiper. Breeds along streams and lakes of Mogollon Plateau and just below, rarer northward, but *not* breeding on San Francisco Mountain. Common transient statewide, and winters more or less commonly along Colorado River and in central Arizona, north irregularly to Camp Verde.

112. ***Tringa solitaria*** Wilson. Solitary Sandpiper. Fairly common transient statewide, except in Yuma area, where it is uncommon; spring records scarce in west and north. One seen near Yuma in winter, Dec. 26, 1950.

113. *Catoptrophorus semipalmatus* (Gmelin). Willet. Fairly common migrant statewide, particularly in fall and along the Colorado River. Sometimes occurs in June.

114. *Totanus melanoleucus* (Gmelin). Greater Yellowlegs. Rather common migrant in small numbers statewide; winters along Colorado River and in south, when rare except in the Yuma and possibly lower Gila River regions. Sometimes occurs in June.

115. *Totanus flavipes* (Gmelin). Lesser Yellowlegs. Common fall transient in north and (formerly only?) extreme southeast. Rather uncommon migrant over Arizona, otherwise, particularly scarce westward.

116. *Calidris canutus* (Linnaeus). Knot. Casual; one at Topock, *July 23, 1952,* and one about three miles above Imperial Dam, Oct. 2, 1959. Also, one on California side of Havasu Lake, Aug. 9, 1950.

117. *Erolia melanotos* (Vieillot). Pectoral Sandpiper. Uncommon fall migrant, recorded chiefly in the lower Colorado Valley. One winter record, Martinez Lake above Imperial Dam, Dec. 30, 1957. Unrecorded as yet in the northeast.

118. *Erolia bairdii* (Coues). Baird's Sandpiper. Fall transient, uncommon in west to common in east and north. Spring records, cited in A.O.U. Check-list, are based on specimens of *Ereunetes mauri.*

119. *Erolia minutilla* (Vieillot). Least Sandpiper. Common to abundant transient statewide, especially in fall; winters more or less commonly from western to central Arizona.

120. *Erolia alpina* (Linnaeus). Dunlin; Red-backed Sandpiper. Fairly common transient and rare winter visitant in western Arizona, scarcer to east and unrecorded in the north.

121. *Limnodromus griseus* (Gmelin). Short-billed Dowitcher. Rare or casual transient. Only one definite record, near Springerville, *Sept. 21, 1937.* Possibly not as scarce as this absence of specimens implies, but certainly far outnumbered normally by the following species. There is no basis for statement (A.O.U. Check-list) that it winters all through the southwestern United States.

122. *Limnodromus scolopaceus* (Say). Long-billed Dowitcher. Common to fairly common migrant statewide; winters sparingly in southern Arizona and along the Colorado River, generally in flocks. Rare in late June and early July, but no specimens between *May 15* and *August 13.*

123. *Micropalama himantopus* (Bonaparte). Stilt Sandpiper. Unknown except from the Salt River Valley near Phoenix, where it is an uncommon transient, once found in numbers in *April,* once found in June, near Palo Verde, *June 12, 1955,* and in September, at Blue Point, *Sept. 24, 1949.* There is one sight record from the San Pedro Valley.

124. ***Ereunetes pusillus*** (Linnaeus). Semipalmated Sandpiper. Known definitely only from one specimen: near Sasabe, *Apr. 23, 1957.* However, may really be less rare due to difficulty in separating it from other species. Statement (A.O.U. Check-list) that it has been found as a rare transient, both spring and fall, in the intermountain region rests largely or wholly on misidentification.

125. ***Ereunetes mauri*** Cabanis. Western Sandpiper. Common transient statewide, especially in fall. Probably winters rarely and irregularly; records of two near Peoria, *Jan. 15, 1958,* two at Picacho Reservoir, *Dec. 22, 1958,* one on California side of Colorado River about 32 miles above Imperial Dam, Jan. 29, 1957, and two at California end of Imperial Dam, Dec. 23, 1960.

126. ***Limosa fedoa*** (Linnaeus). Marbled Godwit. Rare to uncommon spring transient and fairly common fall transient, with most records from the lower Colorado River and the lakes of the Flagstaff region. One winter record, San Pedro River east of Santa Catalina Mountains, Jan. 27, 1886.

127. ***Crocethia alba*** (Pallas). Sanderling. Rare to uncommon tranisent along the lower Colorado River.

Family Recurvirostridae: Avocets and Stilts

128. ***Recurvirostra americana*** Gmelin. American Avocet. Fairly common migrant statewide; sometimes occurs in large flocks in fall along lower Colorado River, where one occasionally winters. Occasional in late June in central and western Arizona, but not breeding.

129. ***Himantopus mexicanus*** (Müller). Black-necked Stilt. Breeds locally along lower Colorado River and probably at Picacho Reservoir; formerly nested on San Pedro River. Otherwise a fairly common migrant in southern and western Arizona, common in fall along the Colorado River. Three winter records: north of Aztec on Gila River, Dec. 11, 1921, west of Gila Bend, *Feb. 17, 1940,* and near Wikieup, *Nov. 12, 1960.*

Family Phalaropodidae: Phalaropes

130. ***Phalaropus fulicarius*** (Linnaeus). Red Phalarope. Rare fall transient, records chiefly from along lower Colorado River. One winter record: tank south of Sierrita Mountains, Pima County, *Jan. 3, 1959.*

131. ***Steganopus tricolor*** Vieillot. Wilson's Phalarope. Common to fairly common transient statewide, an occasional bird lingering through June; birds seen after mid-June (including at least 105 at Picacho Reservoir *June 20, 1952*) may be returning fall migrants.

132. ***Lobipes lobatus*** (Linnaeus). Northern Phalarope. Common fall transient in northern and western Arizona only, scarce some years in the west; rare in spring, and in the southeast.

Family Stercorariidae: Jaegers and Skuas

133. *Stercorarius pomarinus* (Temminck). Pomarine Jaeger. Two records: near Flagstaff, *late October or early November, 1927,* and lower Havasu Lake, *Sept. 26, 1950.*

134. *Stercorarius parasiticus* (Linnaeus). Parasitic Jaeger. Two records: Bill Williams Delta, *Oct. 13, 1947,* and lower Havasu Lake, *Sept. 19, 1953* (the latter may be *Stercorarius longicaudus,* Long-tailed Jaeger).

Family Laridae: Gulls and Terns

135. *Larus glaucescens* Naumann. Glaucous-winged Gull. Two records: lower Havasu Lake, *Feb. 24, 1954,* and Colorado River about eight miles above Imperial Dam, *Nov. 17, 1956.*

136. *Larus occidentalis* Audubon. Western Gull. One record: Parker Dam, *Dec. 12, 1946.* Some of the large immature gulls observed in winter occasionally along the Colorado River may be Westerns, but it is believed they are chiefly Herring Gulls.

137. *Larus argentatus* (Pontoppidan). Herring Gull. Uncommon winter visitant along lower Colorado River, especially Havasu Lake; known elsewhere only from near Tucson (two specimens). *Larus thayeri,* Thayer's Gull, is included here, following A.O.U. Check-list, though sometimes thought to be a distinct species (one record, Havasu Lake, *Dec. 13, 1946*).

138. *Larus californicus* Lawrence. California Gull, Fairly common to uncommon migrant, chiefly in spring, along the lower Colorado River; elsewhere only one valid record, Long Lake, Coconino County, *Nov. 20, 1932.* One summer record: lower Havasu Lake, *July 28, 1954.*

139. *Larus delawarensis* Ord. Ring-billed Gull. Common to abundant transient and winter visitor along the Colorado River, fairly common on Salt River lakes. Also a common transient on larger lakes of the Flagstaff and St. Johns regions, but rather rare elsewhere.

140. *Larus pipixcan* Wagler. Franklin's Gull. Rather rare migrant along Colorado River; also record of two near Joseph City, Apr. 26, 1948. Only two specimens: Imperial Dam, *Oct. 11, 1956,* and Colorado River about 15 miles above Imperial Dam, *Oct. 24, 1959.*

141. *Larus atricilla* Linnaeus. Laughing Gull. Only one record, near Imperial Dam, *Sept. 3, 1960.* An old sight record from the Colorado River has never been otherwise substantiated, and is very dubious.

142. *Larus philadelphia* (Ord). Bonaparte's Gull. Uncommon transient and rare winter visitor along the lower Colorado River; scattered records in spring and fall from northern Arizona, and five records from southern Arizona (including one in winter at Tucson, found recently dead, *Jan. 2, 1950*).

[***Larus heermanni*** Cassin. Heermann's Gull. Hypothetical. One record, Colorado River about 30 miles above Yuma, Nov. 13, 1955. Other sight records, such as one from the California side of Havasu Lake, July 13, 1948, should be disregarded.]

143. ***Xema sabini*** (Sabine). Sabine's Gull. Rare fall transient along Colorado River (not "accidental" as termed by A.O.U. Check-list); also flock of seven at Martinez Lake above Yuma, Apr. 13, 1956, plus one Apr. 27, 1956, and one seen east of "new" Tacna, Yuma Co., Apr. 16, 1960. Elsewhere in state, only a few fall records, from Tucson, Mormon Lake, Big Lake, Grand Canyon, and Kingman, where one taken from flock *Oct. 1, 1959.* Fall records for Colorado Valley are *Sept. 8,* Oct. 3, and Oct. 9, 1948; Sept. 23 (three birds) and Oct. 2, 1949; Oct. 14, 1950; Sept. 27, 1952; Sept. 30, 1955 (photographed); and *Nov. 4, 1960.*

144. ***Gelochelidon nilotica*** (Gmelin). Gull-billed Tern. One specimen from two seen, Colorado River about 33 miles above Imperial Dam, *May 24, 1959.*

145. ***Sterna forsteri*** Nuttall. Forster's Tern. Common to fairly common transient in lower Colorado Valley, elsewhere rare.

146. ***Sterna hirundo*** Linnaeus. Common Tern. Uncommon fall transient along lower Colorado River and in northern Arizona; also a few fall records from elsewhere in the state, including one near Peoria, *Oct. 5, 1956.* Due to difficulty of separating this species from the foregoing, it could occur more often than suspected.

147. ***Sterna albifrons*** Pallas. Least Tern. Casual: one specimen, lower Havasu Lake, *June 18, 1953.* Also, flock of four or five at Mormon Lake, Sept. 4 to 9, 1933, and one above Imperial Dam, July 30, 1959. A published record for the White Mountains (Big Lake) really pertains to *Chlidonias niger.*

148. ***Hydroprogne caspia*** (Pallas). Caspian Tern. Fairly common transient along lower Colorado River; elsewhere, seen at Sullivan Lake, north of Prescott, Apr. 30, 1936, and fall sight records from Mormon and Long Lakes in Flagstaff area.

149. ***Chlidonias niger*** (Linnaeus). Black Tern. Common transient along Colorado River, and fairly common over rest of state, especially in fall. No positive breeding record, though records span the summer.

Family Columbidae: Pigeons and Doves

150. ***Columba fasciata*** Say. Band-tailed Pigeon. Common summer resident of mountains, northwestern to southeastern Arizona. Rare and irregular in extensive forests of pure ponderosa pine, and in winter in southeast. Casual transient on desert, straggling to Big Sandy Valley, Ajo and

Growler Mountains, and even Yuma. Statement that it winters at Prescott and on Verde River (A.O.U. Check-list) not based on any valid record known to us.

151. *Zenaida asiatica* (Linnaeus). White-winged Dove. Abundant summer resident in southern and central Arizona, and along Colorado River below Davis Dam. Mostly rare and irregular in winter. Has probably increased greatly since 1870's, but destruction of much nesting habitat is now threatened.

152. *Zenaidura macroura* (Linnaeus). Mourning Dove. Common resident in valleys of southern, central, and most of western Arizona, abundant in farmlands. Common summer resident in the north, where it is generally rare and local in winter.

153. *Columbigallina passerina* (Linnaeus). Ground Dove. Resident in better-watered valleys of southern and central Arizona, especially common in summer. Occasionally seen to north, even to Flagstaff (*Oct. 30, 1931*) and Grand Canyon Village (Oct. 22-23, 1930); also in southern Nevada. Rare or absent some winters. Occasionally wanders into foothills near resident range.

154. *Scardafella inca* (Lesson). Inca Dove. Common resident of cities and towns of central and central-southern Arizona; also has been found at Bisbee, Yuma, Parker, and the California side of Parker Dam, where not yet permanently established. Has straggled to various mountains in southeast. Probably absent from state prior to 1870.

Family Psittacidae: Parrots and Macaws

155. *Rhynchopsitta pachyrhyncha* (Swainson). Thick-billed Parrot. Formerly erratic visitant, mainly in winter, to southeastern Arizona, chiefly in mountains. The last reliable reports were in 1922 and 1935, and the last major flight in *1917-18*; rumors to 1945. Ranged north centuries ago to the Verde Valley.

Family Cuculidae: Cuckoos, Roadrunners, and Anis

156. *Coccyzus americanus* (Linnaeus). Yellow-billed Cuckoo. Breeds along main rivers of Sonoran zones, chiefly of southern and central Arizona; rare transient on desert and in towns. Transients pass north through southern Arizona in last third of June.

157. *Geococcyx californianus* (Lesson). Roadrunner. Common resident in Sonoran zones (chiefly Lower Sonoran) of southern, central and western Arizona; scarce in north-central part, and rare in northeast, where still no nesting records. Wanderers appear in high mountains occasionally.

158. *Crotophaga sulcirostris* Swainson. Groove-billed Ani. Casual fall straggler into southeastern Arizona; only one presently confirmable record, Pinery Canyon, Chiricahua Mountains, on or about *Oct. 16, 1928,* though two others are cited by the A.O.U. Check-list.

Family Tytonidae: Barn Owls

159. *Tyto alba* (Scopoli). Barn Owl. Fairly common resident throughout most of the open Sonoran zones except in the deserts of the southwest. Apparently local, but ranging into lower Transition Zone, in the northeast; no winter records north of Needles, California, Salome, and Camp Verde.

Family Strigidae: Typical Owls

160. *Otus asio* (Linnaeus). Common Screech-owl; "Screech Owl"; Mexican Screech-owl; Saguaro Screech-owl. Common resident throughout open Sonoran Zones except in northeast, where it is scarce.

161. *Otus trichopsis* (Wagler). Spotted Screech-owl; Whiskered Owl. Resident in the southeast, where rather common in some of the border mountain ranges in heavy Upper Sonoran Zone woodlands.

162. *Otus flammeolus* (Kaup). Flammulated Screech-owl; Flammulated Owl; Scops Owl. Common summer resident in most of the Transition Zone, particularly where oaks are present. Apparently not rare as a transient in lowlands of south in spring, and recorded once at bottom of Grand Canyon. One fall record possibly from non-breeding area: Burton, southern Navajo County, *Sept. 2, 1949.* One winter record, *Feb. 16, 1949,* at Phoenix.

163. *Bubo virginianus* (Gmelin). Great Horned Owl. Fairly common resident statewide, except in densest forests.

164. *Glaucidium gnoma* Wagler. Mountain Pygmy-owl; "Pygmy Owl"; Rocky Mountain Pygmy-owl. Resident, but generally uncommon, of Transition and locally Upper Sonoran Zone woods west to central Arizona: Pajaritos (Atascosa) and Santa Catalina mountains, Prescott, and near Ash Fork, and once (nest) on South Rim of Grand Canyon. Has reached lowest edge of Upper Sonoran Zone twice, *Jan. 21* and *Apr. 7, 1950.*

165. *Glaucidium brasilianum* (Gmelin). Ferruginous Pygmy Owl; Ferruginous Owl. Local and generally sparse resident of Lower Sonoran Zone in central-southern and central Arizona, from Saguaro Lake, Superior, and Tucson west rarely to desert ranges of southern Yuma County.

166. *Micrathene whitneyi* (Cooper). Elf Owl. Common summer resident in southern Arizona lowlands, scarcer north of Gila Valley in central Arizona, ranging up into live oak belt in most of the mountains. Scarce to rare in Colorado Valley below Davis Dam, commoner on Big Sandy River. Records from Prescott, *June 20, 1892,* lower Oak Creek, *Aug. 29, 1956,* and Chloride, Mohave County, *May 27, 1960.*

167. *Speotyto cunicularia* (Molina). Burrowing Owl. Rare and local summer resident in Sonoran zones grass- and farmlands, except in farm areas about Phoenix and Yuma, where fairly common. Formerly commoner. Also rare fall transient, mainly in northeast and on southwest deserts; one record on Mogollon Plateau (White Mountains, October). Somewhat commoner in southern and central Arizona in winter, when one northern Arizona record: Springerville, Jan. 8, 1959.

168. *Strix occidentalis* (Xantus). Spotted Owl. Uncommon resident of the heavily forested mountains and high mesas. Rare in lowlands not far from mountains, perhaps chiefly in winter. Nested near Tucson, *1872*. Absent from most or all areas inhabited by *Bubo*.

169. *Asio otus* (Linnaeus). Long-eared Owl. Rather rare winter resident generally but of statewide occurrence. Even scarcer in summer, but may breed at almost any point in eastern and central Arizona; has nested as far southwest as Bates Well, Organ Pipe Cactus National Monument, *June, 1932*. Found breeding commonly, recently, in Tombstone region. Birds taken, freshly dead, from upper Havasu Lake, *July 9, 1948* (!) and from highway a few miles west in California, *May 7, 1952*. Winter flocks may number up to about 50 birds.

170. *Asio flammeus* (Pontoppidan). Short-eared Owl. Generally rare winter visitant in open grassland, marshes, swales, and fields of southern and western Arizona, also records for extreme desert: one south of Mohawk, Yuma County, Nov. 27, 1942; several in Yuma and western Pima counties, 1959-60. Rare in migration in the north. A report of numbers on the Colorado River in September is doubtless erroneous.

171. *Aegolius acadicus* (Gmelin). Saw-whet Owl. Resident, perhaps fairly common but not often detected, of mountains of eastern, central, and perhaps northwestern Arizona, also possibly winter visitant to some extent in same areas. Rare and irregular in lowlands of central-southern and west-central Arizona in winter. Status poorly known.

Family Caprimulgidae: Goatsuckers

172. *Caprimulgus vociferus* Wilson. Whip-poor will; Stephens' Whip-poor-will. Common summer resident of Transition and adjacent Upper Sonoran Zones of southeastern and (in recent years) central Arizona; ranges west to Pajaritos Mountains, and northwest less commonly (no specimen) to the Hualapai Mountains. Two lowland records: near Roosevelt, *Nov. 4, 1952*, and Tucson, Oct. 6, 1953.

173. *Caprimulgus ridgwayi* (Nelson). Preste-me-tu-cuchillo; Ridgway's Whip-poor-will; Buff-collared Nightjar. One record only, Guadalupe Canyon in extreme southeastern corner of Arizona, *May 12, 1960*. It is believed to breed here and in adjacent New Mexico, where it was first found in *1958*.

174. *Phalaenoptilus nuttallii* (Audubon). Poor-will; Nuttall's Poor-will. Common in summer about hills and rocky outcrops of Sonoran zones (rarely higher) throughout state, less common in southwest. Winters (in small numbers as far as known) in southern and western Arizona.

175. *Chordeiles minor* (Forster). Common Nighthawk; Western Nighthawk. Common summer resident, and abundant in migration, in open parts of Upper Sonoran Zone and above, in central and northern Arizona; less common and very local summer resident in mountains of northwestern, central-southern, and southeastern Arizona. Extremely rare transient (no specimens) in Tucson Valley, unknown elsewhere in Lower Sonoran Zone closer than Indian Springs, Nevada.

176. *Chordeiles acutipennis* (Hermann). Lesser Nighthawk; Texas Nighthawk. Common to abundant summer resident in Lower Sonoran Zone of southern and western Arizona, except in most of Yuma County away from the Colorado and Gila Valleys. Two winter records: Phoenix, *Dec. 27, 1897,* and southwest corner of Papago Indian Reservation, Jan. 6, 1940. No record between early January and March.

Family Apodidae: Swifts

[***Cypseloides niger*** (Gmelin). Black Swift. Hypothetical. Status not quite satisfactory. Probably transient in east. One old specimen in American Museum of Natural History labelled "Arizona," without further data. Flock of about 35 seen over pond near Pima, Graham County, May 30, 1953, also one at same place, Aug. 17, 1954; two sight records in fall at Flagstaff, one of which is apparently the basis of statement in A.O.U. Check-list that it migrates through Arizona. Also seen twice by Mearns in 1880's (once on May 6, 1887 atop "Squaw Peak, Verde Mountains").]

177. *Chaetura pelagica* (Linnaeus). Chimney Swift. Only one verifiable Arizona record: non-breeding pair at Tucson, May 30 to *mid-June, 1952.* Many records of genus not determined to species. A transient reported from California side of Colorado River above Yuma, *May 6, 1930.*

178. *Chaetura vauxi* (Townsend). Vaux's Swift. Fairly common but irregular migrant in central-southern and western Arizona, east at least to Santa Catalina and Huachuca Mountains.

179. *Aëronautes saxatilis* (Woodhouse). White-throated Swift. Common breeding bird at cliffs in the mountains and mesas except in southwest, where it nests only at Parker Dam, and perhaps sparingly in the Kofa and Castle Dome mountains of Yuma County. Found in winter irregularly in southern and western Arizona.

Family Trochilidae: Hummingbirds

180. *Calothorax lucifer* (Swainson). Lucifer Hummingbird. Casual: two old records, Fort Bowie, Cochise County, *Aug. 8, 1874,* and "Arizona" (*date?;* old specimen in U. S. National Museum).

181. *Archilochus alexandri* (Bourcier and Mulsant). Black-chinned Hummingbird. Common summer resident in certain deciduous associations of Sonoran zones, including cities; generally absent from deserts. Migrates across the Mogollon Plateau, at least in fall. Scarce in western Arizona after mid-June.

182. *"Calypte" costae* (Bourcier). Costa's Hummingbird. Common breeding bird in the desert regions of central and western Arizona, but *not* in Huachuca Mountains. Disappears almost completely by early July in western Arizona, to reappear in October and spend the winter in western Pima County and in Yuma County from the south side of the Kofa Mountains southward. Young birds remain later on the east side of the Santa Catalina Mountains, to *Aug. 10, 1884.* Aside from the above Yuma County region, it evidently does not normally occur between early August and late January, contrary to the statement (A.O.U. Check-list) that it "winters over most of breeding range," including "Williams River" (*i.e.,* Big Sandy River). Casual at Phoenix, *Dec. 16, 1959,* and at Clifton, *Mar. 9, 1936.*

183. *"Calypte" anna* (Lesson). Anna's Hummingbird. Migrates in rather small numbers into southern Arizona in September and early October and winters there until December and rarely to early March. Has not been found east of Huachuca Mountains, Fort Grant, and Roosevelt Lake, nor along Colorado River farther north than Imperial Dam. Recorded from Hualapai Mountains, *July 19, 1959.*

184. *Selasphorus platycercus* (Swainson). Broad-tailed Hummingbird. Common summer resident throughout boreal and Transition zones, and among deciduous trees along streams in adjacent Upper Sonoran Zone. Migrates uncommonly through lower country between or adjacent to breeding territories, principally in spring; once heard as far west as the Ajo Mountains, Mar. 23, 1947. We have no fall records in the lowlands of northwestern Arizona, and reports in the adjacent Nevada lowlands would seem to require specimen support. Hybrid with *"Calypte" costae* reported from Rincon Mountains.

185. *Selasphorus rufus* (Gmelin). Rufous Hummingbird. Common spring migrant from west slope of Baboquivari Mountains westward south of the Gila River almost to Colorado River, mid-February to *April*; very rare in spring farther east, and casual in the north (one record, Flagstaff, *about Apr. 25, 1952.*) Abundant fall migrant in northern and eastern Arizona, occurring in smaller numbers in central and southwestern Arizona. One substantiated winter record: Tucson, December 1950 to *Jan. 14, 1951*; other recent sight records of this or similar hummingbirds at same place in winter, and specimen from there, *Nov. 11, 1938.*

186. *Selasphorus sasin* (Lesson). Allen's Hummingbird. Rather uncommon early fall transient (chiefly in July) in the mountains of central southern Arizona, east to Mule Mountains (at Bisbee) and Benson. One verified spring record for Heart Tank, Sierra Pinta, Yuma County, *Feb. 18, 1955,* and one southwest of Sonoyta, Sonora, *Feb. 22, 1955.* Owing to confusion with Rufous Hummingbird may be more common than records indicate.

187. *Atthis heloisa* (Lesson and Delattre). Heloise's Hummingbird; Morcom's Hummingbird. Accidental: two female specimens, Ramsey Canyon, Huachuca Mountains, *July 2, 1896.*

188. *Stellula calliope* (Gould). Calliope Hummingbird. Uncommon to rare spring migrant in southwestern Arizona, chiefly from Baboquivari Mountains westward. Common fall transient in mountains of northern and eastern Arizona, occasional in lowlands.

189. *Eugenes fulgens* (Swainson). Rivoli's Hummingbird. Fairly common summer resident in mixed Upper Sonoran and Transition zones of mountains of southeastern Arizona, north to Grahams and Santa Catalinas; possibly occurs northwest to Sierra Ancha, but no specimens. Casual near Tucson, Nov. 11, 1944. A hybrid with *Cynanthus latirostris* taken in Huachuca Mountains.

190. *Lampornis clemenciae* (Lesson). Blue-throated Hummingbird. Uncommon summer resident of moist canyons in the Huachuca and Chiricahua Mountains. Reports from Santa Rita Mountains require verification. Only one specimen record from Santa Catalinas, *May 14, 1884,* and one sight record. No record whatever at "lower elevations," where said to winter by A.O.U. Check-list; a report from "Tucson" is an error.

191. *Amazilia verticalis* (Deppe). Violet-crowned Hummingbird. Found chiefly in the Guadalupe Mountains of extreme southeast, where it breeds. Has been recorded rarely in summer from Sonoita Creek, and the Huachuca and especially Chiricahua mountains.

192. *Hylocharis leucotis* (Vieillot). White-eared Hummingbird. Formerly a rare summer visitant to southeastern mountains, north to Santa Catalinas and Chiricahuas. Only one authenticated record since *1933*, a female taken in Cave Creek Canyon, Chiricahua Mountains, *July 4, 1961.* Supposed winter date listed in A.O.U. Check-list is an error.

193. *Cynanthus latirostris* Swainson. Broad-billed Hummingbird. Common summer resident in mesquite-sycamore associations from the Guadalupe Mountains west along the border, locally, to the Baboquivaris, and north at least to the Santa Catalinas. One record for Chiricahuas, June 6,

1947. One verified winter record, near Tucson, to Dec. 4, 1960. Occurs rarely about Tucson in migration, sometimes remaining for considerable periods.

Family Trogonidae: Trogons

194. *Trogon elegans* Gould. Coppery-tailed Trogon. Uncommon (irregularly?) summer resident of the Huachuca, and, in recent years, the Santa Rita and Chiricahua Mountains; there are old sight records for the Santa Catalinas. One winter record, a young male near Tucson, Jan. 17, 1953.

Family Alcedinidae: Kingfishers

195. *Megaceryle alcyon* (Linnaeus). Belted Kingfisher. Fairly common to common transient wherever there are permanent fish-inhabited waters, and winters where these do not freeze; also sometimes appears at temporary waterholes on desert. Although there are some summer records for central and northern Arizona, there is no good evidence of nesting in the state in the present century.

196. *Chloroceryle americana* (Gmelin). Green Kingfisher; Texas Kingfisher. Rare straggler into Santa Cruz drainage (Tucson and above) and San Pedro Valley (Benson and above) in fall and winter. No authentic record west of Arivaca (*Dec. 23, 1960*); Coues' records for the Colorado River (1865) are very much open to question. Though called "casual" in A.O.U. Check-list, there are specimens from Fairbanks, Nogales, Arivaca, Patagonia (three seen, two taken), and Tucson, almost certain sight records for the latter two places, and sight records for Benson and St. David. Most of the records are from *Oct. 1* to *February*.

Family Picidae: Woodpeckers and Wrynecks

197. *Colaptes auratus* (Linnaeus). "Yellow-shafted" Flicker. Rare transient and winter visitant, recorded in most parts of central and southeastern Arizona; very few northern Arizona records as yet and only two specimens, Cedar Ridge (40 miles north of Cameron), *May 7, 1947*, and near Grand Canyon Village, *Apr. 17, 1957*. Mid-winter records limited to Phoenix and Tucson. The most southwestern record is for the Big Sandy Valley, Mohave County.

198. *Colaptes "cafer"* (Gmelin). "Red-shafted" Flicker. Common summer resident of forested mountains. Winters commonly in areas with trees below, and uncommonly in, the Transition Zone. Intergrades with *C. a. luteus* or *borealis* appear very rare; intergrades with *C. a. mearnsi* occur in upper San Pedro and upper Santa Cruz valleys, and apparently the Mayer and (in present century) Camp Verde regions. All the flickers are regarded as conspecific.

199. ***Colaptes "chrysoïdes"*** (Malherbe). "Gilded" Flicker. Common resident in the wooded Lower Sonoran Zone, including sahuaros, from San Pedro Valley west; scarcer in Yuma County and along the Colorado River, where it ranged 100 years ago up to Fort Mojave.

[***Dryocopus pileatus*** (Linnaeus). Pileated Woodpecker. Hypothetical. Found once, near Point Imperial on North Rim of Grand Canyon, where one was seen Aug. 30, 1935, and workings photographed. The record, however, was not substantiated, though it is cited in A.O.U. Check-list.]

[***Centurus carolinus*** (Linnaeus). Red-bellied Woodpecker. Erroneously reported on the basis of two specimens in a collection, examined, doubtless erroneously labeled.]

200. ***Centurus uropygialis*** Baird. Gila Woodpecker. Common resident throughout Lower Sonoran Zone of southern and western Arizona; rather local in extreme southeast. Winters fairly commonly in adjacent Upper Sonoran Zone, casually reaching the lower edge of Transition Zone. Occurs occasionally in Prescott region. There are winter sight records north to near Pierce's Ferry, Mohave County.

201. ***Melanerpes erythrocephalus*** (Linnaeus). Red-headed Woodpecker. Accidental: one taken *about June, 1894* in the Chiricahua Mountains, and one seen repeatedly in Phoenix from March to May, 1959.

202. ***Melanerpes formicivorus*** (Swainson). Acorn Woodpecker; Ant-eating Woodpecker; Mearns' Woodpecker; California Woodpecker. Common resident among large oaks in mountains throughout Arizona (except extreme north and in Baboquivaris, where rare and local, or perhaps irregular). Straggles from breeding range at all seasons except early spring (but mainly in fall), in some years even to Colorado River and Organ Pipe Cactus National Monument.

203. ***Asyndesmus lewis*** (Gray). Lewis' Woodpecker. Fairly common summer resident of certain Transition Zone parks in San Francisco Mountains, rarer and local northward and eastward. More or less uncommon (in most years) transient and winter visitant in open Upper Sonoran and low Transition woody areas. Occasionally winters in lowlands, commonly so in flight years.

204. ***Sphyrapicus varius*** (Linnaeus). Yellow-bellied Sapsucker; Red-naped Sapsucker; "Red-breasted" Sapsucker. Nests in Canadian Zone of mountains along and north of Mogollon Plateau, and (in some years only?) of the Hualapai Mountains, but very rare except in parts of White Mountains. Common transient throughout state except in open desert areas, where one seen along west side of Sierra Pinta, Yuma

County, Mar. 19, 1958. Common in winter (except in north where very rare) in Sonoran Zone deciduous trees. A hybrid with *S. thyroideus* reported from the Huachuca Mountains.

205. *Sphyrapicus thyroideus* (Cassin). Williamson's Sapsucker; Natalie's Sapsucker. Nests in aspens from Mogollon Rim northward, more or less commonly. Winters in Transition and (sparingly) high Upper Sonoran zones south and west of Mogollon Rim, rarely on Mogollon Plateau, and once at Grand Canyon Village, Feb. 6, 1949. Very rare winter visitant in the Lower Sonoran Zone, usually near mountains but west casually to Colorado River (two sight records).

206. *Dendrocopos villosus* (Linnaeus). Hairy Woodpecker; White-breasted Woodpecker; Chihuahua Woodpecker. Common to fairly common resident of coniferous woodlands. Formerly descended, uncommonly, in winter to adjacent valleys, but no recent record at any distance from conifers.

207. *Dendrocopos pubescens* (Linneaus). Downy Woodpecker; Batchelder's Woodpecker. Sparse resident in deciduous trees of Transition and Canadian zones from White Mountains (where less uncommon), Sierra Ancha, and San Francisco Mountains northward, including (exceptionally) Navajo Indian Reservation. Apparently somewhat less rare in winter, when it reaches Upper Sonoran Zone. Casual in southern Arizona, two records, near Kelvin, Gila County, *April, 1882*, and at Tucson, *March 13, 1954.*

208. *Dendrocopos scalaris* (Wagler). Ladder-backed Woodpecker; Cactus Woodpecker. Common resident throughout Lower Sonoran Zone, as well as in open parts of Upper Sonoran Zone except in northeast.

209. *Dendrocopos nuttallii* (Gambel). Nuttall's Woodpecker. Accidental; one taken at Phoenix, *June* (*or Jan.?*) *24, 1901.*

210. *Dendrocopos "arizonae"* (Hargitt). Arizona Woodpecker. Fairly common resident of live oaks of southeastern Arizona, west and north of Baboquivari, Santa Catalina, and Graham Mountains. Rare in winter in adjacent lowlands.

211. *Picoïdes tridactylus* (Linnaeus). Northern Three-toed Woodpecker; Alpine Three-toed Woodpecker. Uncommon resident in boreal zones, rare in Transition Zone, from White to San Francisco Mountains and on the Kaibab Plateau.

Family Cotingidae: Cotingas

212. *Platypsaris aglaiae* (Lafresnaye). Rose-throated Becard; Xantus' Becard. Local and irregular summer resident along Sonoita Creek, Santa Cruz County; in Guadalupe Mountains in southeastern corner of Arizona; and in Tucson vicinity. A report from Chiricahua Mountains lacks valid basis. Before *1947*, only one record: Huachuca Mountains, *June 20, 1888.*

Family Tyrannidae: Tyrant Flycatchers

213. *Tyrannus tyrannus* (Linnaeus). Eastern Kingbird. Rare summer visitant in northern Arizona. There are also records of single birds seen near Parker, Yuma County, and Marana, Pima County, and one taken at Cottonwood, Verde Valley, all in early September, and in southwestern Utah, Aug. 29. Accidental near Tucson, Mar. 19-20, 1943. No Arizona nest yet recorded.

214. *Tyrannus vociferans* Swainson. Cassin's Kingbird. Common summer resident in Upper Sonoran and highest Lower Sonoran zones, openings in ponderosa pines, and along major streams except Colorado and lower Salt and Gila rivers, where it is a rare transient. Casual in winter recently at Tucson (no mid-winter specimen for state). The "January" record for Nogales is an error.

215. *Tyrannus verticalis* Say. Western Kingbird; Arkansas Kingbird. Common summer resident in various open associations below the Transition Zone (local in northern Arizona), and fairly common transient (especially in fall) elsewhere in unforested areas, up to the Transition Zone. No authentic record between October and March.

216. *Tyrannus melancholicus* (Vieillot). Tropical Kingbird; Couch's Kingbird; West Mexican Kingbird. Nests near Tucson in open cottonwoods; also recently elsewhere in Santa Cruz and possibly San Pedro valleys and along Salt River east of Phoenix. Wanders rarely in late summer and fall to Colorado River, north to Topock. One casually at Imperial-Riverside county line, California side of Colorado River, Mar. 22, 1957.

217. *Tyrannus crassirostris* Swainson. Thick-billed Kingbird. Local summer resident in riparian trees of Guadalupe Mountains in extreme southeastern Arizona, where it was first found *June 4, 1958,* and in the Patagonia area, where at least four pairs nested in 1961. Breeds within 15 miles of Nogales in Sonora, but has not been found on the Arizona side in that area.

218. *Muscivora forficata* (Gmelin). Scissor-tailed Flycatcher. Rare or casual summer visitant, always singly, chiefly to southeast; specimens from Kayenta, *July 8, 1934,* near Willcox, *Aug. 7, 1953,* and Pomerene, *May 8, 1957.* Also records from Saguaro Lake, July 12, 1935; near Phoenix on Black Canyon Highway, May 18, 1958; and near Tucson, Sept. 17, 1960. A few others of the genus not identified as to species.

219. *Myiodynastes luteiventris* Sclater. Sulphur-bellied Flycatcher. Fairly common summer resident of sycamore-walnut canyons of Santa Rita, Huachuca, and Chiracahua, rarely Santa Catalina and Graham, mountains. One lowland record reported, north of Nogales, *May 27, 1971.*

220. *Myiarchus crinitus* (Linnaeus). Great Crested Flycatcher; Crested Flycatcher. Casual; one taken in Huachuca Mountains, *June 3, 1901.*

221. ***Myiarchus tyrannulus*** (Müller). Wied's Crested Flycatcher; Arizona Crested Flycatcher. Common summer resident of sahuaro, cottonwood, and sycamore associations north to central Arizona, much less common toward Colorado River, where it is found locally north to Topock area, and at Yuma has apparently nested in the city.

222. ***Myiarchus cinerascens*** (Lawrence). Ash-throated Flycatcher. Common summer resident throughout all but the densely wooded parts of the Sonoran zones, rarely straggling higher. Winters in mountains of Yuma County, and sparsely along lower Colorado River, and east to Phoenix and Casa Grande areas, rarely to Tucson.

223. ***Myiarchus nuttingi*** Ridgway. Nutting's Flycatcher; "Pale-throated Flycatcher." Casual: one record, near Roosevelt Lake, *Jan. 8, 1952.*

224. ***Myiarchus "tuberculifer"*** (Lafresnaye and D'Orbigny). Olivaceous Flycatcher. Common local summer resident of denser live oaks and higher riparian associations in Sonoran zones from Guadalupe Mountains west beyond Nogales, and north to the Graham Mountains. There are specimens from the south end of the White Mountains, *July, 1951;* The Baboquivari Mountains, *May 27, 1931*; and even Sells, Papago Indian Reservation, *July 8, 1918.* It is reported in spring migration below the Huachuca Mountains.

225. ***Sayornis phoebe*** (Latham). Eastern Phoebe. Rare fall transient and winter visitor in southern Arizona, chiefly in the southeast but recorded west to the Colorado River. Ten specimens, north to Salt River and west to Bill Williams Delta. Not "accidental," as stated in A.O.U. Check-list.

226. ***Sayornis nigricans*** (Swainson). Black Phoebe. Breeds commonly along permanent streams and canals of central and southeastern Arizona, and uncommonly and locally in the bottom of the Grand Canyon, in western Arizona, on the north slope of the White Mountains region, and even up into the Transition Zone. Winters at water throughout Lower Sonoran Zone valleys, including the Colorado Valley, sparingly so east of San Pedro Valley. On migration apt to occur at any water-hole, even in the northeastern part of the state and up into the ponderosa pine zone. Not "resident" as stated in A.O.U. Check-list.

227. ***Sayornis saya*** (Bonaparte). Say's Phoebe. Fairly common breeding bird about cliffs and structures throughout less densely wooded parts of the Sonoran zones, and locally higher. Winters south and west of the Mogollon Rim in Sonoran zones, sparingly north of Springerville, the west side of the Navajo Indian Reservation, and probably inside the Grand Canyon. During fall migration may be found on Kaibab and Mogollon Plateaus away from breeding areas. A post-breeding migration carries most of the birds out of the southwestern part of the state, where virtually absent in July and August.

228. *Empidonax fulvifrons* (Giraud). Buff-breasted Flycatcher. Rare summer resident in Arizona, more common and widespread formerly; at present known regularly only from low Transition Zone of the Huachuca Mountains, but probably still occurs in southern parts of White Mountains region. There are records from as far north as Prescott (*May 9, 1865*) and Fort Apache, also from the Santa Catalina, Rincon, Santa Rita, Chiricahua, and Patagonia mountains. On spring migration occurs at foot of mountains, even to Sonoita Creek (*Mar. 30, 1927*) and west to Pajaritos Mountains west of Nogales (*Apr. 13, 1947*).

229. *Empidonax wrightii* Baird. Gray Flycatcher. Common summer resident of pinyon-juniper areas from Mogollon Rim northward; also westward to at least Juniper (and probably Aquarius) Mountains, and south to Fort Apache, southern Navajo County. Winters sparingly in mesquite associations, usually near water, in southern and central Arizona, north to Topock, Salome, and Wickenburg, and east to the Patagonia region (casually to Chiricahua Mountains). Common transient in more open parts of state, except in southwest where uncommon.

230. *Empidonax oberholseri* Phillips. Dusky Flycatcher; Wright's Flycatcher. Locally common summer resident of Canadian Zone willows of White and (very locally) San Francisco mountains, and possibly also the Kaibab Plateau. Winters casually along lower Colorado and Salt rivers, and in Chiricahua Mountains; more regularly near Tucson and Patagonia, where not exceptional. Though said by A.O.U. Check-list to winter "casually" in Arizona, nine specimens have been taken near Tucson in the period between Nov. 20 and Mar. 10, in eight different winters; four in two years at Patagonia; four near Phoenix; one at Parker; one near Needles, California; and one or two in the Chiricahua Mountains. An uncommon migrant throughout wooded areas of the state except along lower Salt, Gila, and Colorado rivers, where apparently absent; common in extreme southeast.

231. *Empidonax hammondii* (Xantus). Hammond's Flycatcher. Common throughout state during migrations. Winters regularly in small numbers on Sonoita Creek near Patagonia, casually near Tucson, Phoenix, and in Chiricahua Mountains, and accidentally at Salome.

232. *Empidonax "minimus"* (Baird and Baird). Least Flycatcher. A rare fall migrant in western Arizona: three immature specimens from the Big Sandy Valley, *Oct. 20, 1951*; one from the Tule Mountains, extreme southwestern Arizona, *Sept. 29, 1956*. Also one from Boulder City in Nevada, *Sept. 6, 1950*. Unrecorded farther east.

233. *Empidonax difficilis* Baird. Western Flycatcher. Common summer resident in boreal zones throughout southeastern Arizona, and in the northeast west to the Kayenta region and San Francisco Mountains; breeds down into more shady parts of Transition Zone and possibly even into oak-sycamore associations locally. Absent from northwestern and central Arizona and the Grand Canyon. Common transient in southwestern and central Arizona, rare in north (fall records only) and extreme east, and unknown in northeast. Singles found in Bill Williams Delta, *Dec. 13, 1950,* and Jan. 5, 1951, and there are two winter records for central-northern Sonora near the Arizona line. Records of transients nearly span the summer.

234. *Empidonax flaviventris* (Baird and Baird). Yellow-bellied Flycatcher. Accidental, one specimen: Tucson, *Sept. 22, 1956.*

235. *Empidonax virescens* (Vieillot). Acadian Flycatcher. Accidental, one specimen: Tucson, *May 24, 1886.*

236. *Empidonax traillii* (Audubon). Traill's Flycatcher; Little Flycatcher; Alder Flycatcher. Breeds locally in dense willow association and buttonbush swamps of Sonoran zones (very locally in Transition Zone) throughout state. Transient throughout state, being especially common in southwest, possibly only casual (fall only) in the extreme northeast; migrants pass through south chiefly in first half of June (extremes *May 15,* in California, to *June 23*) and through most of August and September.

237. *Contopus "pertinax* Cabanis and Heine." Coues' Flycatcher. Common summer resident of Transition Zone of southeastern and central Arizona, northwestward almost to Prescott and sparingly northward to Baker's Butte (Mogollon Rim) and White Mountains; on migration found in adjacent Upper Sonoran Zone, and exceptionally in lower Whetstone and Baboquivari Mountains and even along Salt and San Pedro Rivers. Three winter records: Patagonia, Wickenburg, and upper Santa Catalina Mountains.

238. *Contopus virens* (Linnaeus). Eastern Wood Pewee. One immature female specimen: Tucson, *Oct. 7, 1953.* An immature from the Chiricahua Mountains, *Sept. 16, 1956,* is possibly of this species.

239. *Contopus sordidulus* Sclater. Western Wood Pewee. Common summer resident almost throughout the Transition Zone, and in heavy pinyon stands, in walnut-ash-sycamore associations, and very locally in cottonwoods of upper part of Lower Sonoran Zone. Breeds west to Baboquivari (locally) and Hualapai mountains. Common transient throughout state.

240. "*Nuttallornis borealis* (Swainson)." Olive-sided Flycatcher. Fairly common summer resident in extensive boreal zone forests, less common in adjacent ponderosa pine-Gambel oak association, of northeastern Arizona, west to Kaibab Plateau and south to the entire Mogollon Plateau and (irregularly?) the Sierra Ancha. Fairly common migrant over entire state, less common toward west side.

241. *Pyrocephalus rubinus* (Boddaert). Vermilion Flycatcher. Common to abundant summer resident in mesquites, willows, and cottonwoods (always near water in the lower valleys) in southern and central Arizona, but rather local along the Salt and Colorado River valleys. Also nests locally in sycamore-ash-cottonwood associations. Casual in pine-oak-juniper (Natanes Plateau, Gila County, Apr. 15, 1937). Winters in moister valleys of most of breeding range, but sparingly or not at all in extreme southeast except at San Bernardino Ranch. Uncommon transient at or near water in southwestern deserts.

242. *Camptostoma imberbe* Sclater. Beardless Flycatcher. Fairly common summer resident in cottonwood, heavy mesquite, and even sycamore-live oak-mesquite associations, north to Gila River (at mouth of San Pedro River), and from New Mexican border west to west side of Baboquivari Mountains. Winters in the vicinity of Tucson, on the east side of the Baboquivaris, and along the lower San Pedro River.

Family Alaudidae: Larks

243. *Eremophila alpestris* (Linnaeus). Horned Lark. Nests in open grasslands throughout the state, and possibly in some farmlands, thus absent during the breeding season in open areas without grass. Winters commonly in same areas, also in fields, parks on the plateaus, and sometimes on barren shores of rivers and lakes. Absent from all brushy or wooded areas. Casual above timberline in White Mountains.

Family Hirundinidae: Swallows

244. *Tachycineta thalassina* (Swainson). Violet-green Swallow. Common summer resident in most of Transition and Canadian zones, and locally in Upper Sonoran Zone cliffs near water in northeast. There are isolated Lower Sonoran colonies in Havasupai Canyon, near Camp Verde, and along the Colorado River (nests on California side) at Parker Dam. Common throughout state on migration, but not in extreme west in fall. The early return of some birds in late January and February gives an erroneous impression of wintering; as a matter of fact, there are no December specimens, and only a few valid sight records of one or two individuals each. The only November specimen is an immature taken from a cat in Tucson *"about the middle of November" 1944.* Thus the statement (A.O.U. Check-list) that it winters upon the Colorado River to Needles, California, is erroneous.

245. *Iridoprocne bicolor* (Vieillot). Tree Swallow. Winters commonly along the lower Colorado River, and occasionally eastward as far as Picacho Reservoir in Pinal County; also, probably, in extreme southeastern Arizona and the Big Sandy Valley. Generally distributed during migrations, usually along streams or at lakes and ponds. It has been recorded throughout June at Topock, but is usually absent from Arizona from *Mid-May* to *early-July*.

246. *Riparia riparia* (Linnaeus). Bank swallow. Fairly common to rare transient at lakes, ponds, irrigated fields, etc., throughout Arizona.

247. *Stelgidopteryx ruficollis* (Vieillot). Rough-winged Swallow. Common summer resident in dirt banks of streams throughout Sonoran zones of state, irregular on Mogollon Plateau (at Flagstaff). Rather common transient at and along waters. Winters rarely along Colorado River, where its spring arrival is in *January;* one was seen at Picacho Reservoir, Nov. 19, 1958.

248. *Hirundo rustica* Linnaeus. Barn Swallow. Local summer resident in eastern Arizona towns and ranches from Sonoita, San Pedro, and Sulphur Springs valleys north to Show Low, Snowflake, and Holbrook and east (also breeds in Nogales, Sonora); irregularly in Santa Cruz Valley and at Mayer, Yavapai County. Common in migration at and along waters and over fields throughout state. Has been found in winter in small numbers at Picacho Reservoir, Pinal County, and probably wintered once in extreme lower Colorado Valley.

249. *Petrochelidon pyrrhonota* (Vieillot). Cliff Swallow. Nesting colonies are found almost throughout the state in vicinity of water, mainly on cliffs, dams, bridges, and culverts, occasionally on buildings. Common transient at rivers, lakes, fields, etc., statewide, and southward migration apparently starting in late June. Occasionally seen in February and early March along lower Colorado River, and one found at Tucson, *Feb. 10, 1948.*

250. *Progne subis* (Linnaeus). Purple Martin. Breeds in Transition Zone of open parts of the entire Mogollon Plateau region, even to such areas as the William region, the Hualapai Mountains (where rare), Mount Trumbull, the Natanes Plateau, the Sierra Ancha, and the Prescott region; also in the Chiricahua Mountains, but absent from the other mountains of southern Arizona, the Grand Canyon, and the northeast. Also, in sahuaro associations of central southern and central Arizona, west to the Ajo Mountains and north to near Picacho (Pinal County), Florence, Roosevelt Lake, and lower San Pedro Valley, but only nesting at one point (Arivaca) in this region outside of the sahuaros. Very rare outside of breeding ranges, but strays occasionally to the lower Colorado River on migration.

Family Corvidae: Jays, Magpies, and Crows

251. *Perisoreus canadensis* (Linnaeus). Gray Jay; Canada Jay; Rocky Mountain Jay. Common resident of the firs and spruces of the White Mountains.

252. ***Cyanocitta stelleri*** (Gmelin). Steller's Jay; Long-crested Jay. Common resident of pine, fir, and spruce forests throughout state; also may breed very locally in high Upper Sonoran Zone. Winter range extends into oaks of southern Arizona. Large flights occur in some years into southern and western lowlands, when the birds may be found over the entire desert, even to near Yuma.

253. ***"Aphelocoma" coerulescens*** (Bosc). Scrub Jay; Woodhouse's Jay. Common resident of dense Upper Sonoran brush and woodlands, west and south to Hualapai and Baboquivari mountains. In most, if not all, winters a few descend to Lower Sonoran brush, orchards, and trees, mostly along streams, of central Arizona, and to brush of Harquahala and Kofa Mountains; and in occasional years it is quite common in such lowland areas throughout the state. Migrates across the Mogollon Plateau, at least in some autumns.

254. ***"Aphelocoma" ultramarina*** (Bonaparte). Arizona Jay; Mexican Jay; Gray-breasted Jay. Common resident of live oaks in southeastern and central Arizona, west to Baboquivari and Santa Catalina Mountains, and north sparingly to a number of points below the Mogollon Rim as far as the northwest corner of Gila County. There are a few records outside the Upper Sonoran Zone, mainly from the lowlands about Tucson.

255. ***"Cissilopha" san-blasiana*** (Lafresnaye). San Blas Jay; "Black and Blue Jay." Accidental. Two taken from flock of about eight near Tucson, *Dec. 19, 1937*; one in same locality, *Dec. 19, 1938;* and one *Jan. 15, 1939.*

256. ***Pica pica*** (Linnaeus). Black-billed Magpie; American Magpie. Only two specimens: near Winslow, *Dec. 8, 1853,* and Rio Puerco at Navajo Springs, Apache County, *June 27, 1873.* (Also a fresh skin found at Tuba City, *1938.*) Prior to about 1885 said to be common in parts of northeastern Arizona. Now found only in San Juan River drainage of extreme northeast, where one was seen at Tees-Nos-Pas, Sept. 19, 1936. No verified record for Mogollon Plateau or southward.

257. ***Corvus corax*** Linnaeus. Common Raven; American Raven. Fairly common resident almost throughout open parts of Arizona wherever nesting cliffs are available. Rare in Flagstaff area, lower Colorado Valley below Ehrenberg, and as a summer resident near Phoenix. Large congregations occur in northern and eastern Arizona at times.

258. ***Corvus cryptoleucus*** Couch. White-necked Raven. Common summer resident of yucca-mesquite-grasslands association of southeastern Arizona north to Safford area, ranging uncommonly into lower, less grassy brush; limits of nesting range uncertain, but may extend to west edge of Papago Indian Reservation. Winters over most of breeding range, although large numbers appear to leave in conspicuous migrations in mid-November.

259. *Corvus brachyrhynchos* Brehm. Common Crow. Locally common resident of open parts of entire Mogollon Plateau, and down Salt River into Lower Sonoran Zone; also in the Chuska Mountains and the Defiance Plateau of the Navajo Indian Reservation, and perhaps elsewhere in the north. During winter may be seen in adjacent areas, rarely as far from breeding range as the Big Sandy or even lower Colorado valleys. No specimens as yet south of the Gila Valley, which is apparently its limit of occurrence in any numbers.

260. *Gymnorhinus cyanocephalus* Wied. Piñon Jay. Common resident of juniper-pinyon regions in northern and central Arizona (south possibly to Prescott area) eastward to Natanes Plateau north of San Carlos, and west to at least the Hualapai Indian Reservation and the Mount Trumbull area. Sometimes invades adjacent forests. Wanders erratically in fall and spring (rarer in winter), in some years even to the Mexican border and the lower Colorado River.

261. *Nucifraga columbiana* (Wilson). Clark's Nutcracker. Common resident in boreal zones to timberline in the San Francisco and White Mountains, perhaps also on the Kaibab Plateau; has bred casually on South Rim of Grand Canyon. During fall and winter there are occasional large-scale invasions of other mountains, when it ranges also into lower country, even to the lower Colorado River; following these the birds may linger well into the summer in mountains, and even at Boulder City, Nevada (1951).

Family Paridae: Titmice, Verdins, and Bushtits

262. *Parus atricapillus* Linnaeus. Black-capped Chickadee; Long-tailed Chickadee. The only Arizona specimen is from Betatakin Ruin, Navajo National Monument, *Oct. 23, 1936*, where about 10 were seen in October, 1935. Sight records elsewhere unsubstantiated, but *may* occur in Grand Canyon National Park (both rims?).

263. *Parus sclateri* Kleinschmidt. Mexican Chickadee. Common resident in pine and spruce-fir forests of the Chiricahua Mountains. There is a winter sight record for adjacent, lower Swisshelm Mountains. A specimen labeled "Huachuca Mts." was doubtless actually taken in Chihuahua.

264. *Parus gambeli* Ridgway. Mountain Chickadee. Common resident in pine and spruce-fir forests, locally into pinyon-juniper in northeast, throughout mountains except Hualapais and Mexican border ranges. In winter ranges uncommonly into Upper and rarely Lower Sonoran areas adjacent to its breeding range in northern Arizona, casually to Big Sandy River (1880) and San Pedro River (*1961*).

265. *Parus inornatus* Gambel. Plain Titmouse; Gray Titmouse. Fairly common resident in Upper Sonoran Zone of northwestern, northern, central, and locally southeastern Arizona; west to Mount Trumbull and the

Cerbat, Hualapai, Bradshaw, Graham, and Chiricahua Mountains. Casual at foot of Santa Catalina Mountains, *Nov. 28, 1928,* and in bottom of Grand Canyon (Supai), *Nov. 18, 1912.*

266. *Parus wollweberi* (Bonaparte). Bridled Titmouse. Common resident of Upper Sonoran woodlands of southeastern and central Arizona, north to Mogollon Rim (as at Oak Creek Canyon), and west to Juniper, Weaver, Pinal, and Baboquivari mountains. Nests (irregularly?) in Lower Sonoran Zone cottonwood-willow-mesquite locally, at least in Camp Verde region and at mouth of Aravaipa Creek. Regular winter visitant to willow-cottonwood association along larger streams in southern and central Arizona, west to Tempe and Sacaton.

267. *Auriparus flaviceps* (Sundevall). Verdin. Fairly common resident of whole Lower Sonoran Zone except bottom of Grand Canyon; also among mesquites in country otherwise mainly Upper Sonoran. Has evidently increased with the spread of mesquite.

268. *Psaltriparus minimus* (Townsend). Bushtit; Common Bushtit; Lead-colored Bushtit. Rather common resident of Upper Sonoran woodlands, and even chaparral and scrub oaks, throughout Arizona, including Harquahala Mountains. Wandering flocks are sometimes found in other zones, July to March. Has been taken in mountains just east of Colorado River, but never in Colorado Valley.

Family Sittidae: Nuthatches

269. *Sitta carolinensis* Latham. White-breasted Nuthatch; Rocky Mountain Nuthatch. Rather common resident throughout Transition and lower Canadian zones, also locally among larger trees of the Upper Sonoran Zone (though no nesting record for Baboquivari Mountains); also locally in riparian Lower Sonoran Zone (Santa Cruz Valley). Fairly regular, August to early April, in nearby Upper Sonoran Zone and cottonwoods of Lower Sonoran Zone, in major flight years even to lower Colorado River (Bill Williams Delta, *Nov. 10, 1950,* and about 18 miles above Imperial Dam, *Nov. 26, 1961*).

270. *Sitta canadensis* Linnaeus. Red-breasted Nuthatch. Resident in all, or nearly all, of boreal zones. During the fall sometimes found in Transition and Sonoran zones, usually in large trees but casually in desert shrubs, even ranging west to the Colorado River in flight years; most regular as a fall transient in mountains and mesas of the north, west to Hualapai Mountains. One summer record from Tuba City, *July 2, 1936.*

271. *Sitta "pygmaea"* Vigors. "Pygmy" Nuthatch; Black-eared Nuthatch. Abundant resident in ponderosa pines throughout Arizona, and to some extent in adjacent heavy pinyon-juniper. Wanders to timberline, and once to bottom of Grand Canyon (sight records). One specimen from Lower Sonoran Zone: Yuma, *Sept. 30, 1902* (now lost).

Family Certhiidae: Creepers

272. *Certhia familiaris* Linnaeus. Brown Creeper; Rocky Mountain Creeper; Mexican Creeper. Rather common summer resident of boreal zones and, in the south, Transition Zone, throughout Arizona, wintering uncommonly through Transition Zone (possibly higher) and Upper Sonoran woodlands. Usually rare winter visitant in large trees along rivers, particularly in Camp Verde and Tucson areas (where not uncommon in some winters), and west occasionally to the lower Colorado River.

Family Cinclidae: Dippers

273. *Cinclus mexicanus* Swainson. Dipper; Water Ouzel. Fairly common resident along the few clear, swift, permanent mountain streams of Arizona, mostly along the southern rim of the Mogollon Plateau from Oak Creek east to Black River and Eagle Creek in the White Mountains region, and in the bottom of the Grand Canyon. It is also found sparingly to rarely in the rougher parts of the Santa Catalina, Graham, and Chuska Mountains, and the Sierra Ancha. Rarely wanders to other mountin streams, as in eastern Mohave County near Fort Rock, Nov. 7, 1959. Two records from Huachuca Mountains, *May 28, 1903,* and Aug. (date?) 1902; also more frequently from Chiricahua Mountains, *March 20, 1881,* late February, 1917, and *recently.*

Family Troglodytidae: Wrens

274. *Troglodytes aëdon* Vieillot. House Wren; Parkman's Wren. Common summer resident in dense brush and fallen trees from Transition Zone to timberline, south and west to the Chiricahua, Santa Catalina, Bradshaw, and Hualapai Mountains. Winters commonly in better-vegetated areas of Lower Sonoran Zone of southwestern Arizona and the lower Colorado Valley east to the Phoenix region, Tucson, and Patagonia, casually farther east (even in upper Chiricahua Mountains!). In migration common in southwestern Arizona, uncommon northeastward to almost rare (?) in northeastern Arizona. Integrades with following "species" in Rincon Mountains and perhaps farther south and east.

275. *Troglodytes "brunneicollis"* Sclater. "Brown-throated" Wren; "Apache" Wren. A name applied to those wrens that breed commonly in the Transition and Canadian zones of the Huachuca and Santa Rita Mountains; conspecific with the House Wren. Statements that it is a permanent resident (A.O.U. Check-list) lack factual basis.

276. *Troglodytes troglodytes* (Linnaeus). Winter Wren. Local and usually rare winter resident, generally in the densest brush of the more permanent streams, of Transition and adjacent zones in various parts of the state; unrecorded from a number of mountain ranges. A very few high Lower Sonoran Zone records. One record from the lower Colorado Valley, Parker, *Nov. 1, 1953.* The number of records, particularly on

Oak Creek, where the most casual visits have already produced two specimens, does not justify the statement (A.O.U. Check-list) that it is merely "casual" in Arizona. Other specimens (besides the straggler from Parker) are from "Grand Canyon" and White, Huachuca, and Santa Catalina mountains.

277. ***Thryomanes bewickii*** (Audubon). Bewick's Wren; Baird's Wren. Common resident in Upper Sonoran brush and woodland south and west of the Salt and Verde valleys; summer resident locally and generally uncommonly in pinyon-juniper zone over the rest of the state, and in mesquite-willow-cottonwood association along the higher parts of some rivers. Migrates across the Mogollon Plateau. Winters throughout that part of its breeding range that lies south and west of the Mogollon Plateau; also among dense weeds and brush of the Lower Sonoran Zone, as well as on the Hualapai Indian Reservation, the west side of the Navajo Indian Reservation, and even at Springerville. Recently has become rare in extreme western Arizona, where up to 1950 it was a common winter resident; more common again in fall of 1960.

278. ***Campylorhynchus brunneicapillus*** (Lafresnaye). Cactus Wren. Common resident almost throughout the Lower Sonoran Zone, especially in cholla cactus, but also in open mesquite and shade-trees in towns; lacking only at bottom of the Grand Canyon and in the Verde Valley, where no confirmed report.

279. ***Telmatodytes palustris*** (Wilson). Long-billed Marsh Wren; Western Marsh Wren. Local resident along the lower Colorado River, and in marshes of the lower Salt River. Common winter resident and migrant at reed-grown ponds and canals, except possibly in the northeast, less common in winter at frozen marshes on and near Mogollon Plateau.

280. ***Catherpes mexicanus*** (Swainson). Cañon Wren. Rather common resident about cliffs, hills, and adjacent buildings and even high dirt river-banks in Sonoran zones throughout Arizona, but quite uncommon in southwest and along lower Colorado River. Ranges exceptionally into low Transition Zone.

281. ***Salpinctes obsoletus*** (Say). Rock Wren. Common summer resident in open rocky situations from timberline and above down to Upper Sonoran Zone generally, though locally scarce recently (*e.g.*, Hualapai Mountains, San Francisco Peaks); less common in Lower Sonoran Zone (but in numbers some seasons to the low desert mountain ranges of Yuma County and in the Colorado Valley as far south as Parker Dam). Winters commonly in open, broken areas of the Sonoran zones of southern Arizona, but mostly absent then in northeastern Arizona east of Tuba City area and north of Holbrook.

Family Mimidae: Mockingbirds and Thrashers

282. ***Mimus polyglottos*** (Linnaeus). Mockingbird. Rather common summer resident in the less densely wooded parts of the Sonoran zones that afford thorn-brush, less common except in wettest years west of the Baboquivari Mountains, the mouth of the Salt River, and the Kingman region. Common resident of Maricopa County cities and Yuma. Winters commonly in most of southwestern Arizona west of Altar Valley and south from Roosevelt Lake, Wickenburg, and Davis Dam, and in upper Gila Valley; less common in and near Tucson, and exceptionally in southeastern Arizona and in lower parts of Grand Canyon region. Migrant and wanderer on Mogollon Plateau. Only a straggler in open pine-grass (Transition Zone) in Arizona, though common there locally in northern Sonora. There are peculiar seasonal fluctuations in the numbers of Mockingbirds in southern Arizona not yet fully understood.

283. ***Dumetella carolinensis*** (Linnaeus). Catbird. Locally a fairly common summer resident in dense willow-brush association (Upper Sonoran Zone) in the Springerville region, and probably in Oak Creek Canyon recently; possibly on the west side of the Chuska Mountains. There are records from Cow Springs, Coconino County, Apr. 1934, and Lee's Ferry, Aug. 25, 1909. Casual in the Chiricahua Mountains, *Sept. 16, 1956.*

284. ***Toxostoma rufum*** (Linnaeus). Brown Thrasher. Rare fall and winter visitant to Sonoran zones of southeastern Arizona, chiefly in recent years, with records extending west and northwest to Prescott, Phoenix, and Tucson (including four specimens, October to January, but none from northwest of Tucson). Also taken in Zion National Park, Utah.

285. ***Toxostoma bendirei*** (Coues). Bendire's Thrasher. Rather common to local summer resident of the open parts of the Sonoran valleys almost statewide, except for Sulphur Springs Valley (where breeding is unproven), the plains north of Williams, and the immediate Colorado Valley. Found especially where stretches of open ground meet tall dense bushes and/or cholla cacti. Rare, and possibly only a migrant, in the Sulphur Springs Valley. Winters sparingly from the lower Salt River south to Tucson, but not farther west than the Phoenix region and possibly the Ajo Mountains, nor farther east than Tucson; October to December records are all from a narrow strip from central Arizona southeast to the Tucson region. Migrants apparently return to Tucson in January.

286. ***Toxostoma curvirostre*** (Swainson). Curve-billed Thrasher; Palmer's Thrasher. Very common resident of cholla cactus association, rather common in other dense thorny brush and even in towns, of the Lower Sonoran Zone west to the Growler and Kofa Mountains and the Big Sandy River; occasional west of Growler Mountains to Cabeza Prieta

Mountains. Local in extreme southeast. Rarely straggles outside its breeding area, even to the Colorado River, Camp Verde, and Fort Apache, and (sight records) southeastern Nevada.

287. *Toxostoma lecontei* Lawrence. Le Conte's Thrasher. Uncommon, usually very local, summer (*not* everywhere permanent) resident in the open creosote bush deserts of extreme western and southwestern Arizona, east (at least formerly) in the Gila Valley to the Phoenix-Florence-Picacho Peak region and to east of Ventana Ranch in northern Papago Indian Reservation. Permanent resident in Yuma County south of Gila River and east of Gila Mountains, east to the southwestern part of the Organ Pipe Cactus National Monument.

288. *Toxostoma "dorsale"* Henry. Crissal Thrasher. Common resident in dense, tall brush along rivers and larger washes of Lower Sonoran Zone, locally in dense broad-leafed Upper Sonoran chaparral, south and west of the Mogollon and Kaibab plateaus; but absent west of Growler Mountains and Papago Well in western Pima County. Uncommon fall (at least) migrant in northern Arizona (!): 2 taken *Sept. 22, 1932* from south of Grand Canyon Village; also several records from the lower Little Colorado Valley, and taken at Shumway, *Sept. 3, 1953.*

289. *Oreoscoptes montanus* (Townsend). Sage Thrasher. Fairly common summer resident of sagebrush areas of northeastern and possibly northwestern Arizona, south possibly to Springerville (*July 5, 1936*). Fairly common migrant on open plains, not so common in extreme southwest. Rather common winter resident in sparse brush of Sonoran zones plains of southern Arizona east to at least the San Pedro Valley but less common along lower Colorado River; is also found in winter in the lower Little Colorado Valley. Movements poorly understood. Rare migrant across higher plateaus. Casual (juvenile) specimen from near Gleeson, Cochise County, *June 4, 1940,* and one seen at Flagstaff, Dec. 11, 1938.

Family Turdidae: Thrushes, Solitaires, and Bluebirds

290. *Turdus migratorius* Linnaeus. American Robin. Common summer resident of openings in Transition and boreal zones, and locally in moist Upper Sonoran woodland along the main canyons, throughout Arizona except Hualapai Mountains; found recently in summer in Cerbat Mountains of Mohave County. Summer Lower Sonoran records: Parker, Aug. 16, 1953, and near Kingman, *June 12, 1959.* Winters in Sonoran zones, especially in berry-bearing Upper Sonoran Woodland and in large towns, somewhat irregularly in mistletoe-mesquite association, and uncommonly in Transition Zone.

291. *Turdus rufo-palliatus* Lafresnaye. Rufous-backed Robin. Casual straggler from Mexico: one near Nogales, *Dec. 18, 1960.*

292. "*Ixoreus*" *naevius* (Gmelin). Varied Thrush. Two specimens, near Tucson, *Feb. 5, 1956* and *Mar. 22, 1958.*

293. "*Hylocichla*" *guttata* (Pallas). Hermit Thrush. Common summer resident of boreal zones down very locally even into dense second-growth live oak woods (in Santa Rita Mountains). Rather common winter resident in Sonoran zones of southern and central Arizona (perhaps occasionally to bottom of Grand Canyon), especially in dense Upper Sonoran woods and brush and in shady parts of major river valleys. During migrations found even in the most arid desert mountain ranges, where also irregular in winter. Transient, but rare in most or all non-breeding areas, in northern Arizona.

294. "*Hylocichla*" *ustulata* (Nuttall). Swainson's Thrush; Russet-backed Thrush; Olive-backed Thrush. Rare summer resident of the fir forests of the San Francisco Mountains. Rather common spring migrant through southern and western Arizona from Huachuca Mountains west, north to Kingman region, uncommon farther east and only two May records in the north. Rare and more restricted in the state in fall, when limited mostly to the southern border, chiefly in the mountains, even to Reserve, New Mexico.

295. "*Hylocichla*" *minima* (Lafresnaye). Gray-cheeked Thrush. One specimen: Cave Creek, Chiricahua Mountains, *Sept. 11, 1932.*

296. "*Hylocichla*" *fuscescens* (Stephens). Veery; Willow Thrush. Doubtless breeds in the willow association of the Upper Sonoran Zone southwest of Springerville, where taken *July 3 and 4, 1936.* Otherwise no authenticated record.

297. *Sialia sialis* (Linnaeus). Eastern Bluebird; Azure Bluebird. Rare resident, probably quite local, in the live oaks and nearby pines from the Huachuca Mountains west to beyond Nogales; recent sight records in Chiricahua Mountains also. Rare or casual in winter in the Tucson vicinity, where specimens taken *Feb. 12, 1915* and where seen Jan. 19 to 21, 1952, and at Patagonia, *Nov. 24, 1950.*

298. *Sialia mexicana* Swainson. Western Bluebird; Chestnut-backed Bluebird; Mexican Bluebird. Common summer resident in open Transition and lower Canadian zones (and, in northeastern Arizona, among the larger trees in Upper Sonoran woodlands) west to the Santa Catalina, Bradshaw, and Hualapai mountains, and Mount Trumbull; now scarcer in the southern mountains. Winters abundantly in berry-bearing Upper Sonoran woodland, uncommonly in open Transition Zone woods, and somewhat irregularly in farmlands and on the desert wherever mistletoe occurs.

299. *Sialia currucoides* (Bechstein). Mountain Bluebird. Common summer resident of the open parts of northern Arizona from pinyon-juniper woodland up to timberline, south to the entire Mogollon Plateau region and west to Ash Fork and beyond Mount Trumbull. Winters abundantly in more open berry-bearing parts of Upper Sonoran Zone, uncommonly in Transition Zone openings, and commonly in most years south of its breeding range in farmlands, grasslands, and open berry-bearing woods and brush.

300. *Myadestes townsendi* (Audubon). Townsend's Solitaire. Rather common summer resident in high Transition and boreal zones of northern Arizona, west to the Kaibab Plateau and San Francisco Mountains and south to White Mountains. Migrant and post-breeding wanderer in other parts of Transition Zone. Winters commonly in berry-bearing parts of Upper Sonoran Zone, rarely but more or less regularly in the better-wooded Lower Sonoran valleys and canyons, and even at times in the desert mountains of extreme southwestern Arizona, remaining occasionally to early June. One summer sight record in Chiricahua Mountains.

Family Sylviidae: Old World Warblers, Gnatcatchers, and Kinglets

301. *Polioptila caerulea* (Linnaeus). Blue-gray Gnatcatcher; Western Gnatcatcher. Rather common summer resident in open woodlands and chaparral of the Upper Sonoran Zone, west to the Ajo, Castle Dome, Kofa, and Hualapai or even Mohave mountains, and locally in higher parts of adjacent mesquite associations, at least in the west. Common winter resident of wooded river valleys of Lower Sonoran Zone, north and east to at least the Tucson area and the Salt and Big Sandy rivers, and west to the Colorado River. Rare fall transient in higher forests (a few July and August records); common transient in western deserts north of Gila River.

302. *Polioptila melanura* Lawrence. Black-tailed Gnatcatcher; Plumbeous Gnatcatcher. Fairly common resident of open brush in the Lower Sonoran Zone, commoner westward and rather local east of the San Pedro Valley; not found in Verde Valley or the bottom of the Grand Canyon, nor of course anywhere in the northeast.

303. *Regulus satrapa* Lichtenstein. Golden-crowned Kinglet. Uncommon resident in boreal zones from Santa Catalina and Chiricahua Mountains north and northwestward to Kaibab Plateau. Fairly common fall and winter resident in some of the higher forests of northern Arizona, especially White Mountains, reaching lower Transition Zone. Occasional winter visitant in Lower Sonoran woodlands, chiefly in the Tucson vicinity, but casually west to near Phoenix, Wickenburg (flock!), and even to the Colorado River (23 miles north of Yuma, Nov. 21, 1957), and to the Sierra Pinta of Yuma County (*Dec. 27, 1957*)!

304. *Regulus calendula* (Linnaeus). Ruby-crowned Kinglet. Common summer resident of the more extensive boreal zone forests of northern and locally southeastern Arizona, west and south to the Kaibab and entire Mogollon plateaus and the Graham, Santa Catalina (very locally, 1959), and Chiricahua mountains. Common transient throughout state. Winters commonly in Lower Sonoran Zone, more sparingly in Upper Sonoran Zone south and west of the Mogollon Plateau, and rarely onto, or perhaps even north of, that plateau.

Family Motacillidae: Wagtails and Pipits

305. *Anthus spinoletta* (Linnaeus). Water Pipit; American Pipit. Breeds above timberline in the San Francisco and White mountains. Generally distributed about water, and even occurs on open plains, in Sonoran and Transition zones in migration. Common winter resident at streams, ponds, irrigated fields, etc., throughout the Lower Sonoran Zone, rare to uncommon in winter in open Upper Sonoran Zone.

306. *Anthus spragueii* (Audubon). Sprague's Pipit. Rare in fall and early spring in fields and grasslands south and west of the Mogollon Plateau; only seven records (all of single birds) altogether, extending from Sulphur Springs Valley to the Colorado River, the northernmost near Wikieup in the Big Sandy Valley. One winter specimen: Tule Desert, Yuma County, *Dec. 30, 1958.*

Family Bombycillidae: Waxwings

307. *Bombycilla garrula* (Linnaeus). Bohemian Waxwing. A rare but perhaps fairly regular winter visitant to northern Arizona; *extremely* rare south of the extreme north, but to be looked for almost anywhere in years of great flights. The only four specimens, remarkably enough, are all from the Colorado River: Fort Mohave, *Jan. 10, 1861* (*not* 1871); Willow Beach below Hoover Dam (Arizona side), *Apr. 30, 1938* (apparently now lost); and Davis Dam (two), *Mar. 6, 1959.*

308. *Bombycilla cedrorum* Vieillot. Cedar Waxwing. An erratic winter visitant in Sonoran zones south and west of Mogollon Plateau, often abundant in May in valleys and canyons, with occasional birds or flocks remaining even into June. Occurs as a migrant elsewhere in the state, but not known to remain all winter.

Family Ptilogonatidae: Silky Flycatchers

309. *Phainopepla nitens* (Swainson). Phainopepla. Present in different areas in varying roles, seasonally; probably few, if any, individuals are sedentary or non-migratory. It is also entirely possible that there may be two populations, of different seasonal status, at the same place. Common winter resident in Lower Sonoran Zone of southern Arizona, except Verde Valley and most of the southeast, most common in southwest. Nests almost throughout Lower (and locally Upper) Sonoran Zone in brush containing mistletoe; later it absents itself from western Arizona (east to include Big Sandy Valley and Papago Indian Reservation) from June until late August or, most usually, early October. There is a sprinkling of summer records from most of northern Arizona, and it is said to breed locally in the Snowflake area, and perhaps near Mount Trumbull. Abundant summer resident (and breeds) in parts of southeast, and bred in 1876 near Fort Apache in Upper Sonoran Zone. In occasional winters remains far north and east of usual winter range; Fort Huachuca, *1866-7;* and in *1955-6* to Sedona and the Chiricahua Mountains.

Family Laniidae: Shrikes

310. *Lanius excubitor* (Linnaeus). Northern Shrike. Only four verifiable records: Prescott, *Feb. 6, 1865*; near Flagstaff, *Jan. 7, 1936*; extreme northwestern Coconino County, six miles south of Johnson, Utah, *Nov. 27, 1937*; and near Show Low, *Dec. 21, 1959.* Probably fairly regular in north, but almost no work done there in winter, and very hard to identify.

311. *Lanius ludovicianus* Linnaeus. Loggerhead Shrike; White-rumped Shrike. More or less common summer resident throughout open parts of the state (except brushless grassland) below Transition Zone, rather uncommon (at least in midsummer) along Mexican border west of Baboquivari Mountains. Uncommon transient in Transition Zone. Winters commonly in Lower Sonoran Zone, less commonly in open Upper Sonoran Zone.

Family Sturnidae: Starlings

312. *Sturnus vulgaris* Linnaeus. Starling. Since its first known occurrence in the state in 1946 (near Lupton), the Starling has become established commonly as a breeding resident in the Phoenix area, and to a lesser extent at Kingman, Tucson, and Yuma. It occurs in winter in virtually any settled location in the Sonoran zones, chiefly in farmlands, and abundantly near Kingman and Holbrook (where no recent work has been done in summer). Numbers seen in southern Utah as early as January 1941, when no ornithologists were in northern Arizona; but a report from southern Nevada in August 1938 seems very doubtful.

Family Vireonidae: Vireos

[***Vireo griseus*** (Boddaert). White-eyed Vireo. Erroneously reported from "Cyanthanis" (= Grantham's), Sulphur Springs Valley; the specimen has been examined and proves to be *Empidonax wrightii* Baird = "*Empidonax griseus* Brewster."]

313. *Vireo huttoni* Cassin. Hutton's Vireo; Stephens' Vireo. Rather common summer resident in the denser live oak brush and woods of southeastern Arizona, north sparingly to the Mazatzal Mountains and the Whiteriver-Fort Apache area, and west to the Santa Catalina and Pajaritos Mountains. Fairly common winter resident from Patagonia and the Santa Catalina Mountains west to the Baboquivari and even Quijotoa Mountains, north rarely to Salt River and perhaps the foot of the Natanes Plateau in Gila County (*Nov. 10,* wintering?); mostly in streamside trees and brush and in the lower oaks. Found uncommonly in early fall in the higher conifer forests adjacent to breeding areas. Rare west to the Bill Williams Delta, *Nov. 24, 1953,* the Colorado River above Imperial Dam, *Nov. 13, 1960,* and the Kofa Mountains, *Sept. 25, 1956* and Sept. 29, 1960.

314. *Vireo bellii* Audubon. Bell's Vireo; Arizona Vireo. Common summer resident in dense low brush, especially mesquite association along streams, up to the top of the Lower Sonoran Zone; absent from the bottom of the Grand Canyon, the Mexican border west of the Ajo region, and from the area of the confluence of the Gila and Salt Rivers, but has been found on upper Colorado River (in recent years) up to Toroweap Valley, Grand Canyon National Monument (June, 1953). Breeding range was formerly less discontinuous, as it bred at Phoenix in *1889.* Scarce in recent years along Colorado River. Reduced or exterminated in some regions by Cowbird parasitism. One winter specimen from Topock, Feb. 7 to *Mar. 7, 1951.*

315. *Vireo vicinior* Coues. Gray Vireo. Fairly common summer resident in chaparral-juniper and large junipers of the Upper Sonoran Zone, in southeastern to northwestern Arizona, but absent from the Mexican border ranges west of the Chiricahua Mountains and from the area east of the Grand Canyon and north of the Little Colorado River and St. Johns. A rare migrant in Lower Sonoran Zone of southern and western Arizona. Winters in small numbers in the mountains of Yuma County, and probably westernmost Pima County (not rare near Sonoyta in Sonora, *Nov. 1955*), mostly south of the Gila River; one winter record from near Tucson, *Dec. 31, 1949.* Erroneously reported from Hoover Dam area (specimens are *V. bellii*).

316. *Vireo flavifrons* Vieillot. Yellow-throated Vireo. Three specimens: Chiricahua Mountains, *May 8, 1948*; near Mammoth, *Aug. 17, 1948;* and Bill Williams Delta, *Oct. 10, 1953.* Also one record from Sierra Pinta, Yuma County, June 4, 1956.

317. *Vireo solitarius* (Wilson). Solitary Vireo; Plumbeous Vireo; Cassin's Vireo. Common summer resident in Transition Zone throughout Arizona, and in the heavier vegetation of the Upper Sonoran Zone (locally and irregularly ranging down after breeding into cottonwood association of Lower Sonoran Zone). Common in migration statewide, principally in southwest in spring and mountains in fall. Winters in small numbers in willow-cottonwood association of Lower Sonoran Zone of central-southern and central Arizona, also in towns, dense mesquites, and casually in live oaks.

318. *Vireo olivaceus* (Linnaeus). Red-eyed Vireo. Rather rare transient in deciduous vegetation in northern Arizona; casual in south where there are specimens only from the Huachuca Mountains, supposedly on *May 20, 1895*, and from the Colorado River about 25 miles above Imperial Dam, *May 28, 1956*. Accidental in Tucson, July 6, 1952 and near Tucson May 17, 1958. Also taken in northern Sonora, *July, 1952*.

319. *Vireo philadelphicus* (Cassin). Philadelphia Vireo. Three fall specimens: near Tucson, *Nov. 10, 1939*, and *Oct. 7, 1953*, and King Valley, south of Kofa Mountains, Yuma County, *Oct. 27, 1954*.

320. *Vireo gilvus* (Vieillot). Warbling Vireo. Rather common summer resident in the willows, maples, dense boxelders, and especially aspens from the Hudsonian Zone down, locally and irregularly, to the Upper Sonoran Zone, west to the Santa Catalina and Juniper Mountains; but not in such adjacent Lower Sonoran places as St. Thomas, Nevada, and St. George, Utah, as stated by A.O.U. Check-list. Common in migration throughout the state. Observed once in late February, Baboquivari Mountains.

Family Parulidae: Wood Warblers

321. *Mniotilta varia* (Linnaeus). Black-and-white Warbler. Rare winter resident near Tucson, at least since *1938*. Casual farther west and north; only three other Arizona specimens, Bill Williams Delta, *Feb. 15, 1952*, Wickenburg, *1959*, and Huachuca Mountains; but two others from Boulder City, Nevada. There are sight records from the Salt River at Saguaro Lake and from Parker. Though called "casual" by A.O.U. Check-list, there are now five Tucson specimens, from five different years, plus sight records.

322. *Protonotaria citrea* (Boddaert). Prothonotary Warbler. Two specimens: Tucson, *May 1, 1884*, and Chiricahua Mountains, *Sept. 8, 1924*.

[***Helmitheros vermivorus*** (Gmelin). Worm-eating Warbler. Hypothetical. One report only: Chiricahua Mountains, March 30, 1941.]

323. *Vermivora pinus* (Linneaus). Blue-winged Warbler. Casual: one record, Bill Williams Delta, *Sept. 5, 1952*.

324. *Vermivora peregrina* (Wilson). Tennessee Warbler. One specimen, Chiricahua Mountains, *Apr. 7, 1925*. A few recent sight records in the southeastern mountains on migration.

325. *Vermivora celata* (Say). Orange-crowned Warbler; Lutescent Warbler. Locally a summer resident in dense deciduous cover, chiefly willow thickets, in the Canadian Zone of eastern Arizona (White and Graham mountains, and one locality in the Santa Catalina Mountains), possibly sparingly farther west (Mormon Lake). Common transient statewide. Winters uncommonly from Patagonia and Tucson west and northwest in dense brush; commonly along the Colorado River north to at least Davis Dam. Two midsummer records from southwestern Arizona: Kofa Mountains, June 25 and 26, 1956, and Colorado River 23 miles above Yuma, July 1, 1957.

326. *Vermivora ruficapilla* (Wilson). Nashville Warbler; Calaveras Warbler. Fairly common transient throughout Arizona, especially in large brush and weeds and deciduous trees, particularly southwestward; in spring found east only to San Pedro River, and the Tempe and Prescott areas (one sight record at Flagstaff).

327. *Vermivora "virginiae"* (Baird). "Virginia's" Warbler. Common summer resident of low deciduous brush in the higher mountains throughout Arizona, chiefly in Transition and Canadian zones; also breeds along low rivers north of Mogollon Plateau (Holbrook) in Upper Sonoran Zone. Descends locally (Santa Catalina Mountains) to foothills in July. Uncommon transient (chiefly in spring) in the Lower Sonoran Zone west to the Pajaritos Mountains, Santa Cruz and Big Sandy valleys, and rarely to the Colorado River. Probably a race of *V. ruficapilla.*

328. *Vermivora luciae* (Cooper). Lucy's Warbler. Abundant summer resident in dense mesquite and cottonwood-mesquite associations of the Lower Sonoran Zone, fairly common in ash-walnut-sycamore-live oak association of the Upper Sonoran Zone, in most of southern and central Arizona and along the entire Colorado River. Absent from desert west of Growler Valley and from most of Phoenix area. In recent years has become very scarce along lower Colorado River. Casual on and just north of Mogollon Plateau in migration.

329. *Parula americana* (Linnaeus). Parula Warbler. Rare fall and winter visitant across southern Arizona, with several records from the Tucson area and casual ones west to the Colorado River. Spring records are: near Tucson, *March 26, 1938*; near Roosevelt Lake, May 30, 1952; and Chiricahua Mountains, *May 25, 1957*. Though termed "casual" in A.O.U. Check-list, the period 1938-1953 produced three specimens, and two other individuals seen, near Tucson alone, and it bred in northern Sonora.

330. *Peucedramus taeniatus* (Du Bus). Olive Warbler. Fairly common summer resident, scarcer northwards, in Transition and (locally) Canadian zones of southeastern Arizona, north to the south edge of the Mogollon Plateau and west to Santa Rita and Santa Catalina mountains.

Uncommon winter resident in most, if not all, of its breeding range. One sight record near Tucson.

331. *Dendroica "petechia"* (Linnaeus). Yellow Warbler. More or less common summer resident in willows, cottonwoods, and sometimes sycamores, of Sonoran zones almost throughout state, but peculiarly absent from some areas. Scarce to absent in recent years as a nesting bird along the lower Colorado River. Has bred once in willows of Transition-Canadian zones border. Evidently locally reduced or exterminated by Cowbird parasitism. Common migrant in all parts of Arizona except unbroken woodlands. Casual in winter along the Colorado River (*Parker*, Imperial Dam).

332. *Dendroica magnolia* (Wilson). Magnolia Warbler. Casual transient, four records: Bill Williams Delta, Oct. 5, 1949; Topock, *Nov. 11, 1951*; Tucson, *Nov. 6, 1955*; and Chiricahua Mountains, May 23, 1959.

333. *Dendroica tigrina* (Gmelin). Cape May Warbler. Accidental, one record: *fall of 1875*, probably from Tucson. Also, one found at California end of Laguna Dam on Colorado River, *Sept. 23, 1924*.

334. *Dendroica caerulescens* (Gmelin). Black-throated Blue Warbler. Casual transient: male found dead in Ajo Mountains, *April 30, 1955*; one photographed in color in Chiricahua Mountains, May 5, 1955; one taken (other birds believed seen) in Chiricahua Mountains, *Oct. 17, 1956*; and one seen east of Tucson, Oct. 31, 1959.

335. *Dendroica coronata* (Linnaeus). "Myrtle" Warbler. Rare winter visitant to Lower Sonoran Zone rivers and farms in western and southern Arizona, mostly to west; rare spring transient in eastern, central, and northern Arizona. One fall record in mountains (Graham Mountains, *Oct. 8, 1956*). No *published* record verifiable, however (except A.O.U. Check-list), and sight records generally erroneous. Several specimens intermediate with the following "species" have been taken, and the two are almost certainly conspecific.

336. *Dendroica "auduboni"* (Townsend). "Audubon's" Warbler; "Black-fronted" Warbler. Common summer resident throughout boreal zones (less common in adjacent Transition Zone) except in the Hualapai Mountains and, apparently, central Arizona. Common migrant everywhere. Common winter resident of Lower Sonoran Zone and, locally, deciduous riparian Upper Sonoran Zone of southern, central, and western Arizona, except in driest desert portions; rare in winter in other parts of Upper Sonoran Zone in the south.

337. *Dendroica nigrescens* (Townsend). Black-throated Gray Warbler. Common summer resident in pinyon-juniper woodland, and other dense vegetation of high Upper Sonoran Zone, and apparently also Gambel oak thickets, of eastern Arizona, west to the Baboquivari and Bradshaw mountains, and the Grand Canyon region. Fairly common state-wide on migration. Winters uncommonly in cottonwood-willow and sycamore-mesquite associations in the Phoenix and Tucson areas and in the Baboquivari Mountains. One record from the lower Colorado valley at Yuma, *Feb. 18, 1940*, and also at Bard, California.

338. *Dendroica townsendi* (Townsend). Townsend's Warbler. Transient throughout Arizona, very common at higher altitudes, only fairly common in lowlands; scarce along eastern edge, and unrecorded in northeast in spring. One winter record: Patagonia, *Dec. 3, 1939*. Hybrids with *D. occidentalis* have been taken.

339. *Dendroica virens* (Gmelin). Black-throated Green Warbler. Rare transient; though termed "accidental" in A.O.U. Check-list, there are already seven specimens, from late March to late May, and from late October to early November; localities represented are Huachuca Mountains, Tucson vicinity, Grand Canyon at rim of Toroweap Valley, Ajo Mountains, Parker, and Bill Williams Delta. In addition, reliable sight records come from Tucson, Parker, and Petrified Forest National Monument. Hybrids of *townsendi* × *occidentalis* may closely resemble this species.

340. *Dendroica occidentalis* (Townsend). Hermit Warbler. Abundant fall transient at high altitudes, scarcer at lower elevations, rare in Lower Sonoran Zone. Common spring migrant at all altitudes in southern Arizona west of the San Pedro River, north to Salome and the Mohave Mountains; rare farther east and north to Verde Valley, unrecorded then in the north. One winter record: Big Sandy Valley north of Cane Springs, *Feb. 17, 1958*.

[***Dendroica cerulea*** (Wilson). Cerulean Warbler. One was found dead at Boulder Beach on the Nevada side of Lake Mead, *June 6, 1954*.]

341. *Dendroica graciae* Baird. Grace's Warbler. Common summer resident of Transition Zone throughout the state. One record for Lower Sonoran Zone: St. David, Apr. 19, 1939.

342. *Dendroica pensylvanica* (Linnaeus). Chestnut-sided Warbler. Casual in fall and winter. Two specimens: Wikieup, Big Sandy Valley, *Oct. 4, 1952*, and Phoenix, *Jan. 9, 1955* (the latter lost before prepared). One in Sabino Canyon near Tucson, Oct. 25, 1957.

[***Dendroica striata*** (Forster). Blackpoll Warbler. One seen at California end of Imperial Dam (West Pond), May 15, 1955.]

[***Dendroica discolor*** (Vieillot). Prairie Warbler. Hypothetical. One record near Tucson, Dec. 7 and 8, 1952, the first time on a Tucson Bird Club field-trip.]

[***Dendroica palmarum*** (Gmelin). Palm Warbler. Hypothetical. One seen near Wikieup, Big Sandy Valley, Jan. 29-30, 1951; two at Walnut Grove, Yavapai County, April 29, 1956; and one on California side of Colorado River about nine miles above Imperial Dam, Sept. 22, 1942.]

[***Seiurus aurocapillus*** (Linnaeus). Ovenbird. Hypothetical. One seen at Walnut Grove, Yavapai County, May 4, 1955; one near Patagonia, Santa Cruz County, June 19, 1957; and one at Boulder City, Nevada, June 16, 1954.]

343. ***Seiurus noveboracensis*** (Gmelin). Northern Waterthrush; Grinnell's Waterthrush; Alaska Waterthrush. Rather uncommon transient through whole eastern half of Arizona, and also west locally to Salome. Only two records farther west: Topock, Colorado River, Sept. 21, 1951; and Colorado River about 30 miles above Imperial Dam, Sept. 6, 1958. Also an old record for Nevada (presumably) part of Colorado River, *October*, and reported in southwestern Utah (Zion National Park).

[***Seiurus motacilla*** (Vieillot). Louisiana Waterthrush. Not yet reported from within Arizona, though it probably passes over the southeastern edge in order to reach the Nogales area, where it winters regularly (four specimens, *mid-October* to *mid-December*) but uncommonly within 15 to 27 miles of the city, in Sonora.]

344. ***Oporornis formosus*** (Wilson). Kentucky Warbler. Accidental. One record, Ramsey Canyon, Huachuca Mountains, *May 23, 1959.*

[***Oporornis agilis*** (Wilson). Connecticut Warbler. Reported in A.O.U. Check-list as casual in Cochise County. Basis of this statement is unknown to us.]

345. ***Oporornis tolmiei*** (Townsend). MacGillivray's Warbler. Fairly common summer resident in *Ribes*-willow thickets of the Canadian Zone of the White and (very locally) San Francisco mountains. Common migrant in brush throughout Arizona, except in unbroken forests.

346. ***Geothlypis trichas*** (Linnaeus). Common Yellowthroat; "Yellowthroat." Common summer resident at such reedy marshes as survive in the Sonoran zones of Arizona, and sometimes in dense, tall grass or weedy fields in the southeast. Probably breeds up to lower edge of Transition Zone in White Mountains region. Common migrant at weedy, brushy, and swampy places throughout the more open parts of the state from the Transition Zone (where recorded in fall only) down, even occurring in the most arid desert sections sometimes in spring. Males winter rather commonly (females very rarely) along the lower Colorado River, and locally east as far as Tucson and Safford.

347. ***Icteria virens*** (Linnaeus). Yellow-breasted Chat; Long-tailed Chat. Very common summer resident of dense mesquite-willow-*Baccharis* and arrowweed associations along major rivers and at ponds in Lower Sonoran Zone; fairly common summer resident in deciduous brush along streams and in irrigated lands in Upper Sonoran Zone. Rare transient elsewhere, including the Transition Zone and the arid deserts of southwestern Arizona, and more often garden shrubbery in cities (at least Tucson).

347a. *Euthlypis lachrymosa* (Bonaparte). Fan-tailed Warbler. A female was found in Baker Canyon, Guadalupe Mountains, extreme southeastern Arizona, *May 28, 1961.*

348. *Cardellina rubrifrons* (Giraud). Red-faced Warbler. Common summer resident of Transition and especially Canadian zones, west to the Santa Rita and Santa Catalina mountains, and less commonly north to the Mogollon Plateau and west, in recent years, to the Bradshaw Mountains and the Flagstaff region. Migrant in adjacent Upper Sonoran Zone during cold springs; one lowland record, near Tucson, *April 26, 1940.* Casual in foothills of Santa Catalina Mountains, Feb. 22, 1956.

349. *Wilsonia citrina* (Boddaert). Hooded Warbler. Accidental; one specimen, Patagonia, *July 13, 1947.*

350. *Wilsonia pusilla* (Wilson). Wilson's Warbler; Pileolated Warbler. Common to abundant transient throughout Arizona. There are three winter records: Tucson, *Jan. 27, 1945*; Colorado River about six miles above Imperial Dam, Dec. 27, 1955; and Phoenix, Dec. 30, 1956.

351. *Setophaga ruticilla* (Linnaeus). American Redstart. Probably a rare summer resident on the Little Colorado River near Eagar, where the male of a pair was taken *July 3, 1936,* and a molting female seen, Aug. 12, 1950. In the extreme southeast and in the west it is more or less regular, as single birds, in both spring and fall migrations, remaining into early June. Otherwise a casual transient, principally in the fall. There are recent winter records: one near Tucson, *Jan. 25, 1956*; and Colorado River about six miles above Imperial Dam, *Dec. 20, 1958* (one of two taken), Dec. 23, 1959, and Nov. 4 and Dec. 23, 1960.

352. *Setophaga picta* Swainson. Painted Redstart. Common summer resident of the more heavily wooded parts of the Upper Sonoran Zone (higher live oaks, juniper, pinyon, etc., usually mixed with pines) of all southern and central Arizona, west to the Baboquivari Mountains, and north to the Hualapai Mountains and the Mogollon Rim. Found postbreeding into the Transition and Canadian zones of the same areas and rarely down to lower edge of live oaks. Winters sparingly in the Pajaritos Mountains (and southeast foothills of Baboquivari Mountains? — one specimen), casually elsewhere, in low canyons of the Upper Sonoran Zone of south-central Arizona. Rather rare spring migrant in southeast valleys, casually northwest to Salt and Big Sandy rivers in fall. Accidental in Zion National Park, Utah (sight record only).

Family Ploceidae: Weaver Finches

353. *Passer domesticus* (Linnaeus). House Sparrow; English Sparrow. Abundant resident of cities and towns, and rather common resident of large ranch headquarters, irrigated fields, etc., statewide. Became established in Arizona in the early 1900's, widespread by 1915.

Family Icteridae: Meadowlarks, Blackbirds, and Orioles

354. *Dolichonyx oryzivorus* (Linnaeus). Bobolink. A small breeding colony near Show Low in 1937 is the only one recorded in the state. Otherwise, a rare fall transient in western Arizona; though called "casual" by A.O.U. Check-list, five specimens were taken near Wikieup, *Oct. 2 to 10, 1952*; another was heard on Big Sandy River in late September, 1949, and one was seen in the Bill Williams Delta, Sept. 14, 1954.

355. *Sturnella magna* (Linnaeus). Eastern Meadowlark; Arizona Meadowlark; Texas Meadowlark. Fairly common summer resident of grassy plains and fields in southeastern, central-southern, and northern Arizona. Its centers of abundance are the Altar, Sonoita, San Rafael, and Sulphur Springs valleys, the high prairies of the White Mountains, the grasslands just north of the Mogollon Plateau forests, and the valleys near (especially north of) the Juniper Mountains. It does not extend north of the Little Colorado Valley and the Grand Canyon nor west beyond Coconino County as far as now known. During winter, it is found in varying abundance in southern Arizona west to at least Arlington in the Phoenix region and to the Organ Pipe Cactus National Monument, but is rare, apparently, in the north, except perhaps at Springerville.

356. *Sturnella neglecta* Audubon. Western Meadowlark. Common summer resident in the grassy parts of northern and central Arizona (except most places where *magna* occurs), and locally in irrigated valleys almost throughout Arizona except the southeast, beyond Tucson. Nested at least once in Sulphur Springs Valley in an exceptionally wet year (1941). In wet years it nests also in well-vegetated desert areas of southwestern Arizona. In the northwest it apparently breeds in eastern Mohave County, but not anywhere farther west in the Kingman region; we cannot believe it a "common permanent resident in the . . . desert areas" of adjacent Nevada. Common in grassy and semi-grassy areas in migration. In winter common in grassy parts of the Sonoran zones and farmlands of southern and western Arizona, rather uncommon in grassy or cultivated ponderosa pine openings and in northeastern Arizona farmlands and grasslands.

357. *Xanthocephalus xanthocephalus* (Bonaparte). Yellow-headed Blackbird. Breeds in small colonies at reedy lakes on and north of the Mogollon Plateau, and along the lower Colorado River. Common in migration at marshes and cattle pens, and in smaller numbers at lakes, fields, stock tanks, etc., throughout the state. Winters abundantly, nearly all males, in marshes and adjacent farmlands across southern Arizona.

358. *Agelaius phoeniceus* (Linnaeus). Redwinged Blackbird. Common to abundant resident in marshes and irrigated farmlands of Sonoran zones throughout the state, feeding in the farmlands and roosting in the marshes during winter, when numbers are augmented by birds from more northern areas. Occasionally found in migration at stock tanks, wells, etc., on the desert. Common summer resident at marshy lakes of the Transition Zone and higher in northern Arizona; rare in winter in Transition Zone, and apparently irregular.

359. *Icterus spurius* (Linnaeus). Orchard Oriole. Two specimens, Chiricahua Mountains, *Sept. 2 and 8, 1956.* Also a few recent sight records, from the upper San Pedro Valley east, on migration.

360. *Icterus cucullatus* Swainson. Hooded Oriole. Common summer resident of large mesquite, palm, walnut, and sycamore associations of the Sonoran zones across southern Arizona, north in recent years to below Davis Dam, Kingman, and the south base of the Mogollon Plateau. In migration occurs on deserts of extreme southwestern Arizona. Four winter sight records: Tucson, Dec. 23, 1951, Jan. 26, 1952, and Dec. 1959; and Sierra Pinta, Dec. 27, 1957. Statement that race *nelsoni* winters at Tucson (A.O.U. Check-list) is not based on any specimen known to us.

361. *Icterus pustulatus* (Wagler). Scarlet-headed Oriole. Rare to casual winter visitor (and transient) to the Tucson region, though hardly "accidental," as it has been found *Dec. 19, 1886*, Oct. 30 to *Dec. 26, 1948*, and *March 19 and July 31, 1952*. Also one in Sonora south of Nogales, *Dec. 19, 1956.*

362. *Icterus parisorum* Bonaparte. Scott's Oriole. Common summer resident of the live oak and yucca associations in the mountains of southeastern and central Arizona, and in Joshua trees from Wickenburg region northwest. Smaller numbers breed in the pinyons of most of northern Arizona, and in beargrass (*Nolina*) and *Yucca* in southwestern Arizona, almost to the Colorado River. Occurs as an uncommon migrant in the valleys of southeastern Arizona, and casually (?) at Flagstaff (juveniles, *Aug. 2 and 5, 1947*). May winter irregularly from Baboquivari Mountains west to Organ Pipe Cactus National Monument, but no specimens.

[***Icterus galbula*** (Linnaeus). Baltimore Oriole. Hypothetical: one seen, Colorado River about 18 miles above Imperial Dam, Sept. 22, 1956, and a specimen taken in Sonora south of Nogales, *Oct. 12, 1954.* Regarded as conspecific with *I. "bullockii."*]

363. *Icterus "bullockii"* (Swainson). "Bullock's" Oriole. Breeds commonly in cottonwood-willow association of the Sonoran zones except in parts of Mexican border. On migration ranges throughout the state, especially in Lower Sonoran Zone. There are a very few winter records, including specimens from Parker, *Feb. 1, 1947*, and near Phoenix, *Nov. 21, 1957.*

364. *Euphagus carolinus* (Müller). Rusty Blackbird. Rare fall and winter visitant to at least western, central, and southern Arizona; records include seven specimens (Camp Verde, Picacho Reservoir, Tucson, Globe, Bill Williams Delta, Topock, Tule Well in southern Yuma County).

365. *Euphagus cyanocephalus* (Wagler). Brewer's Blackbird. Common summer resident in the vicinity of willows and in well-watered farmlands on and just below the Mogollon Plateau and in the Springerville region; scarcer in Chuska Mountain region, on the South Rim of the Grand Canyon (irregularly?) and on the Kaibab Plateau; but nowhere farther west. Transient throughout state, generally uncommon except in farmlands. Abundant winter resident, principally in and near farmlands, south and west of the Mogollon Plateau; also winters sparingly at lakes and farms of Mogollon Plateau and northward. Two summer records in Lower Sonoran Zone: California side of Havasu Lake, *June 12, 1947*, and Phoenix, *July 26, 1953*. Also one in low Upper Sonoran (near Tuba City, *July 6, 1936*).

366. *Cassidix mexicanus* (Gmelin). Boat-tailed Grackle; Great-tailed Grackle. Locally resident in recent years in southeastern Arizona as far west as Phoenix; other breeding stations include Safford, San Bernardino Ranch east of Douglas, upper San Pedro Valley, and whole Santa Cruz Valley. There are records from Alamo Canyon, Ajo Mountains, *May 14, 1939*, and Apache Lake, *May 10, 1952*.

367. *Molothrus ater* (Boddaert). Brown-headed Cowbird; Dwarf Cowbird; Nevada Cowbird. Common summer resident of the less densely wooded parts of the Sonoran zones, especially in southern and central Arizona; scarce in pinyon-juniper areas and straggler (breeding?) in adjacent ponderosa pine. Common transient south and west of the Mogollon Plateau. Winters commonly in southern Arizona lowlands where there are livestock concentrations, notably at Parker, Yuma, Willcox, and in the Phoenix and Tucson areas.

368. *Tangavius aeneus* (Wagler). Bronzed Cowbird; Red-eyed Cowbird. Common summer resident (starting early in 20th century) of irrigated areas and sycamore canyons in southern and central Arizona, west to Organ Pipe Cactus National Monument and Wickenburg, north to the Phoenix and Globe regions, and east to the San Pedro Valley and extreme southwestern New Mexico; rarely farther northeast, and west to the Colorado River. Winters, usually in small numbers, at Tucson.

Family Thraupidae: Tanagers

369. *Piranga ludoviciana* (Wilson). Western Tanager; Louisiana Tanager. Common summer resident of the Transition and boreal forests throughout the state except from Prescott region northwest, where lacking. Common, sometimes abundant, transient throughout Arizona, some fall birds arriving in low desert areas early in July, soon after the last spring migrants depart. One winter record, Tucson, *Feb. 11, 1945*.

370. *Piranga olivacea* (Gmelin). Scarlet Tanager. Casual. Three records: Tucson, *May 18, 1884*; Otero Canyon, Baboquivari Mountains, March 2 to 22, 1932; and Wikieup, Big Sandy Valley, *Oct. 19, 1949.*

371. *Piranga flava* (Vieillot). Hepatic Tanager. Common summer resident in dense oaks, fairly common in pines and large pinyons, throughout Arizona, except in north and northwest, where local. Breeding range includes Baboquivari and Hualapai Mountains. Winters rather regularly (but rarely) in Santa Cruz County, from the Sonoita Valley west to the Pajaritos Mountains. Has been recorded on migration in lowlands only five times: near Tucson, *Sept. 11, 1938* (small flock), Oct. 2, 1940, and *Oct. 24, 1958;* in the Castle Dome Mountains of Yuma County, *June 6, 1959*; and Colorado River about 18 miles above Imperial Dam, *Nov. 18, 1960.* A report from the extreme northwest (Virgin River) is unsubstantiated.

372. *Piranga rubra* (Linnaeus). Summer Tanager; Cooper's Tanager. Common summer resident in most of the willow-cottonwood association of the Lower Sonoran Zone, rather uncommon in other broad-leafed cover (sycamores, willow-walnut, etc.) of the Sonoran zones south and west of the Mogollon Plateau. Rare transient, April-May and end of July to September, in the Sonoran zones away from its breeding grounds. In winter has been found in the Phoenix and Tucson areas; also sight records for Patagonia and for near Needles, California.

Family Fringillidae: Grosbeaks, Finches, Sparrows and Buntings

373. *"Richmondena" cardinalis* (Linnaeus). Cardinal. Common resident (reports of altitudinal movements unconfirmed) of the taller and denser Lower Sonoran brush of southern and central Arizona, west to Bates Well on the Organ Pipe Cactus National Monument, the Gila Valley above Gila Bend, and Wickenburg, and north (mainly along rivers) to the foot of the Mogollon Plateau in several places. In recent years, also a small population along the Bill Williams and Big Sandy Rivers that extends to the Colorado River in the Parker Dam-Parker area; and has reached Upper Sonoran Zone (Tonto Natural Bridge, 1955, and Kingman, 1959). Still very local in the southeast. Occasional at Prescott in winter; also a specimen from the Castle Dome Mountains of Yuma County, *Feb. 24, 1956.* The old "Colorado River" records are in error. Has considerably extended its range in past 75 years.

374. *Pyrrhuloxia sinuata* (Bonaparte). Pyrrhuloxia. Rather common summer resident in dense Lower Sonoran brush of southeastern Arizona, from the San Bernardino, middle San Pedro, and Santa Cruz valleys west over most of the Papago Indian Reservation to the Ajo region; in 1958 and 1959 was found westward to the Mohawk and Castle Dome Mountains of Yuma County; still present in the Growler Valley, western Pima

County, in 1960. Winters over same range (except in parts of the San Pedro Valley) and also in the Patagonia area and other areas higher than the breeding range (i.e., Baboquivari Mountains). Has wandered north to Gila Bend, Sacaton, near Superior (regular?), the lower San Pedro Valley, and San Carlos. Coues' report from the Yuma district is to be disregarded.

375. *Pheucticus ludovicianus* (Linnaeus). Rose-breasted Grosbeak. Casual transient or summer visitant to Arizona, where all records are for the southeastern and extreme western parts, from the Chiricahua Mountains west to the Colorado River, mainly in May and June; also one at Kanab, Utah, Apr. 26, 1935. There are but three specimens: Huachuca Mountains, supposedly on *June 29, 1894* and about *Sept. or Oct., 1929,* and Parker, *June 27, 1953.* A winter record, lower Salt River, *Nov. 28, 1958,* is purely accidental. Possibly conspecific with the following "species."

376. *Pheucticus "melanocephalus"* (Swainson). "Black-headed" Grosbeak; Rocky Mountain Grosbeak. Common summer resident of Transition Zone and, in southern Arizona, moist and high Upper Sonoran woodland, especially about creeks and deciduous thickets. Common transient in most parts of state below Canadian Zone, rarely higher, with records practically spanning the summer; "common" at Kingman, June 24-28, 1902!

377. *Guiraca caerulea* (Linnaeus). Blue Grosbeak. Fairly common summer resident of willow, cottonwood, and moist mesquite-farmland associations, etc., of the major valleys of the Sonoran zones. Occasional elsewhere, even on the Mogollon Plateau, including one specimen, White Mountains, *Sept. 12, 1936.* One valid winter record: Parker, *Feb. 18, 1951.*

378. *Passerina cyanea* (Linnaeus). Indigo Bunting. A rare migrant in recent years in eastern and central Arizona, and now nesting at least in Oak Creek and Prescott areas and possibly on Sierra Ancha and along White River. Has crossed with the following "species" at Flagstaff and possibly other localities. Lone males have been seen in southeastern and central Arizona in summer, and one was seen near Tucson, Dec. 6, 1957. Non-breeding females doubtless overlooked, only two records: Chiricahua Mountains, *May 26, 1957*, and Flagstaff, Sept. 15, 1940. No records in the northeast, nor between Flagstaff and southwestern Utah.

379. *Passerina "amoena"* (Say). "Lazuli" Bunting. Rather uncommon summer resident of willow associations of the Sonoran zones of central and northeastern Arizona, south and west to the bottom of the Grand Canyon, Prescott, Camp Verde, and possibly the Mazatzal Mountains and Sierra Ancha. Common transient in brush and tall herbaceous vegetation throughout the less densely wooded parts of the state, records nearly spanning the summer. Winters rarely in southern Arizona from Hereford west and northwest to the Nogales and Phoenix regions (but not west of Santa Cruz Valley), in recent years only and apparently increasing.

380. *Passerina versicolor* (Bonaparte). Varied Bunting; Beautiful Bunting. Locally an uncommon summer resident (breeding in July?) in low thorny thickets of the higher Lower Sonoran Zone in foothill canyons of southern Arizona (west perhaps to the Ajo area), north to the Santa Catalina Mountains, with the Baboquivari Mountains the center of occurrence. There are casual records for near Patagonia, *July 14, 1884;* the east side of the Mohave Mountains near the Colorado River, *Oct. 27, 1949*; the Bill Williams Delta, Sept. 20, 1952; and Wikieup, Oct. 1 to 6, 1952. Also, 15 or more seen at Blythe, California side of Colorado River, in February, *1914*, including two specimens, *Feb. 8 and 9.*

381. *Passerina ciris* (Linnaeus). Painted Bunting. Formerly a fairly common transient (not "casual," as stated by A.O.U. Check-list) in extreme southeastern Arizona, north to the Gila River and west to the Graham Mountains and the Nogales area. Not found in any numbers since *1884*, and seldom recorded since *1914*: one in Pajarito Mountains, *July 27, 1933*, one in Chiricahua Mountains, *Aug. 11, 1956*, and one 10 miles south of St. David, *Aug. 17, 1959.* Still common in Sonora south of Nogales, *Oct. 1954,* but no ornithologist worked in southern Arizona that season.

382. *Spiza americana* (Gmelin). Dickcissel. Uncommon fall transient (not "casual," as implied by A.O.U. Check-list) in Lower Sonoran valleys and farms in the southeast; casual elsewhere, with no records for northeastern corner as yet. All authentic records are between *Aug. 8* and Oct. 16.

383. *Hesperiphona vespertina* (Cooper). Evening Grosbeak. Rather uncommon and erratic summer resident about deciduous vegetation in the Transition Zone of eastern and northern Arizona, west to the Grand Canyon (both rims), the Bradshaw Mountains, the Sierra Ancha, and the Santa Catalina and Santa Rita mountains; possibly also the Hualapai Indian Reservation. Rare and irregular transient and winter visitant in the wooded parts of the Sonoran zones throughout the state, in flight years reaching even Yuma at least once (May 6, 1902). Also generally rare in winter in and near its breeding territory, but common locally in flight years.

384. *Carpodacus purpureus* (Gmelin). Common Purple Finch; "Purple Finch"; California Purple Finch. Irregular fall and winter visitant, usually in small numbers, in central and southern Arizona east to Camp Verde and the Huachuca Mountains (and exceptionally New Mexico). Two taken on South Rim of Grand Canyon, *Dec. 22, 1934*, and one seen at Supai, Sept. 23, 1950.

385. *Carpodacus cassinii* Baird. Cassin's Purple Finch; Cassin's Finch. Common summer resident in boreal forest openings of the Kaibab Plateau, decidedly uncommon and apparently irregular in Transition and Canadian zones elsewhere in northern Arizona, and no definite nesting record. Irregularly abundant winter visitant in high Upper Sonoran and, in southern Arizona, open Transition Zone generally, and very irregularly into Lower Sonoran Zone as at Tucson, Camp Verde, Salome, Kingman, the Big Sandy Valley, the Kofa Mountains, and even near Tule Well, Yuma County, *Nov. 8, 1960*. On the Mogollon Plateau it is chiefly a transient.

386. *Carpodacus mexicanus* (Müller). House Finch; "Linnet." Abundant summer resident in the less dense vegetation of the Sonoran zones, especially about towns and ranches; in recent years has spread to towns and ranches in the Transition Zone. Irregularly reaches even Canadian Zone (Kaibab Plateau). In winter it withdraws into lower valleys, and spreads out commonly to the deserts of western and southwestern Arizona.

387. *Pinicola enucleator* (Linnaeus). Pine Grosbeak. Fairly common and doubtless resident in the boreal forests of the White Mountains. Was present on the Kaibab Plateau June-July, 1929 but not seen there since. There are also two winter records for the South Rim of the Grand Canyon, *Dec. 15, 1950* and *Jan. 6, 1957*. No substantiated record elsewhere.

[***Leucosticte tephrocotis*** (Swainson). Gray-crowned Rosy Finch. Hypothetical. There are a very few sight records, all from the South Rim of the Grand Canyon in winter. None is considered satisfactory; and in any case "*L. atrata*" is regarded as conspecific.]

388. *Leucosticte "atrata"* Ridgway. "Black" Rosy Finch. Specimens were taken at two localities on the South Rim of Grand Canyon, *Nov. 27* and *Dec. 26, 1956*, and there is at least one sight record there (see *L. tephrocotis*).

389. *Spinus pinus* (Wilson). Pine Siskin; Pine Finch. Common summer resident (though hardly any nests recorded) from the Transition Zone to timberline on the Kaibab and Mogollon Plateaus, including the San Francisco and White mountains; rather uncommon to rare as a presumably breeding bird south to the Chiricahua and Santa Rita mountains, and probably to Mt. Trumbull in the northwest. Winters more or less commonly in weedy fields and river valleys almost throughout the state, but rather sparingly in the southwest and along the Colorado River, sometimes appearing in flocks in the Sonoran zones by *Aug. 10* and remaining to *June*.

390. *Spinus tristis* (Linnaeus). American Goldfinch; Pale Goldfinch. Irregularly a rather common winter resident in deciduous trees and weedy fields of the Sonoran zones, sometimes remaining to early June. Transient on the Mogollon Plateau; seen recently there and in Hualapai Mountains, casually, in summer (no summer specimens). Older summer records inside Grand Canyon and on Navajo Mountain (Utah) require substantiation.

391. ***Spinus psaltria*** (Say). Lesser Goldfinch; Green-backed Goldfinch; Arkansas Goldfinch. Fairy common summer resident in deciduous trees and brush (especially willows and cottonwoods) in the Sonoran and (locally) Transition zones throughout Arizona; may also be found locally in live oaks. Irregularly common post-breeding visitor and winter resident in the Sonoran zones, less common in extreme northeast; in late summer visits Canadian Zone (Kaibab Plateau), and may winter irregularly in Transition Zone. Breeding season remarkably prolonged or, more likely, irregular.

392. ***Spinus lawrencei*** (Cassin). Lawrence's Goldfinch. Irregularly common transient and winter visitant in weedy areas and at watering places of the Sonoran zones across southern Arizona, north to the entire Prescott region. In one year at least nested in western Arizona, near Parker in 1952 (nest taken). Specimens in breeding condition taken at Wickenburg, *May 12, 1953,* and in Castle Dome Mountains, *May 6, 1955.* A flock of five seen along Colorado River 20 miles above Imperial Dam, June 2, 1958.

393. ***Loxia curvirostra*** Linnaeus. Red Crossbill; Mexican Crossbill. Irregularly common resident of the more extensive Transition and boreal zones forests, although hardly found breeding as yet. May occur in pinyons in winter. Rare and irregular in the lowlands, where it has been taken below Hoover Dam (Arizona side), *Nov. 14, 1938;* at Tucson, *Nov. 6, 1950* and *Jan. 25, 1956;* and at Parker, *Aug. 23, 1953.*

394. ***"Chlorura" chlorura*** (Audubon). Green-tailed Towhee. Rather common summer resident of low deciduous brush of the Transition and boreal zones of the White and San Francisco mountains and the Kaibab Plateau. Common transient in dense brush throughout Arizona, and winters fairly commonly in low, weedy brush of the Lower Sonoran Zone and adjacent areas of the Upper Sonoran Zone of central and central-southern Arizona; in most winters rather scarce in extreme west, southwest, and southeast.

395. ***Pipilo erythrophthalmus*** (Linnaeus). Rufous-sided Towhee; Spotted Towhee; Spurred Towhee; Mountain Towhee; Nevada Towhee. Common summer resident in dense broad-leafed brush of the Upper Sonoran Zone, ranging locally into the Transition Zone. Winters commonly in the Upper Sonoran foothills of northwestern, central, and southern Arizona, fairly commonly in brushy canyons and river valleys of Lower Sonoran Zone of southeastern Arizona, less commonly westward to the Colorado River, where it is a rather rare transient, and even scarcer in winter in most years. Apparently rare to uncommon in winter in the northeast.

396. *Pipilo fuscus* Swainson. Brown Towhee; Canyon Towhee. Rather common resident in scattered low but dense brush in high Lower Sonoran and low Upper Sonoran zones of most of southern, central, and northwestern Arizona. Restricted to rocky hills and desert mining camps westwardly. Found west to the Black and Mohave mountains of western Mohave County, the Kofa Mountains, and the Ajo region. Island populations may be found on the north side of the White Mountains region from Springerville to near Concho, and along the Puerco River in Apache County. Three sight records along rims of Grand Canyon, from Village west.

397. *Pipilo aberti* Baird. Abert's Towhee. Common resident in dense undergrowth of the willow-cottonwood and large mesquite associations of the main rivers of the Lower Sonoran Zone in southern and western Arizona. The Upper Sonoran Zone report ("Fort Whipple") is an error.

398. *Calamospiza melanocorys* Stejneger. Lark Bunting. Common to abundant winter resident in brushless, weedy, or barren-looking parts of the Lower Sonoran Zone of southeastern Arizona; scarcer and irregular westward but common some years, even to southern Nevada, usually reaching the Colorado River only in fall. Apparently a fairly common transient in eastern Arizona in open Upper Sonoran Zone from Holbrook to Fort Apache, but rather rare farther west or north. A *May 6* record for extreme southwestern Utah is probably exceptional.

399. *Passerculus sandwichensis* (Gmelin). Savannah Sparrow; Large-billed Sparrow. Fairly common summer resident locally at lakes and moist fields on and just north of the Mogollon Plateau in the White Mountains (including Springerville area), also possibly near Flagstaff (and formerly Kayenta?). Common transient at lakes, ponds, marshes, and in fields and level grassy spots throughout the state. Common to abundant in winter in irrigated fields of the lower Colorado Valley and elsewhere in irrigated valleys, grassy swales and plains, along bodies of water, etc., in Lower Sonoran Zone throughout southern and western Arizona. Also found locally and rarely in winter in weedy fields and grassy edges of lakes and ponds in northern Arizona, casually into Transition Zone. Possibly late summer straggler to lakes near Yuma (large-billed form, *rostratus*), authentically recorded only once, *Aug. 15, 1902.*

400. *Ammodramus savannarum* (Gmelin). Grasshopper Sparrow. Fairly common resident in dense grassland and alfalfa fields of the San Rafael and Babocomari valleys, the upper end of the Sonoita Valley, and possibly the Altar Valley-Coyote and Quinlan Mountains region, in southeastern Arizona. Has bred (possibly irregularly) at Fort Grant, and perhaps once (1916) at Camp Verde. Fairly common winter resident in dense grass (usually mixed with low brush) of southeastern Arizona, west to the Papago Indian Reservation; uncommon and irregular farther west, even to the Colorado River where it is mainly a very rare migrant; one winter record years ago at mouth of Bill Williams River (not on Big Sandy River, as stated in A.O.U. Check-list). Only one specimen for northern Arizona, South Rim of Grand Canyon, *June 28, 1941* (probably brought in by tourist's car from elsewhere).

401. *Ammodramus bairdii* (Audubon). Baird's Sparrow. Until about 1878 an abundant transient and doubtless winter resident in the grasslands of southeastern Arizona, north to northern Graham County; until about 1920 still decidedly uncommon winter resident in the Chiricahua and Huachuca Mountains region. Now apparently much rarer, recent records coming only from the Chiricahua, Huachuca, and Santa Rita Mountains areas, south of Bowie (*April*), and the Sonoita Plains. Ranges west (at least in fall) along Mexican border to Sasabe area, but never recorded as far northwest as the Tucson valley. One northern Arizona record, near Eagar, southern Apache County, *Oct. 14, 1934.*

402. *Pooecetes gramineus* (Gmelin). Vesper Sparrow. Fairly common summer resident in dry grasslands from high Upper Sonoran Zone to low Canadian Zone along and north from the Mogollon Plateau. Common migrant in open country generally. Winters commonly in weedy fields and grassy areas in the Lower Sonoran Zone of southern and (locally) central Arizona, northwest to Congress, Salome, and Verde and Big Sandy valleys.

403. *Chondestes grammacus* (Say). Lark Sparrow. Locally a fairly common summer resident in brushy grasslands from high Lower Sonoran Zone to Transition Zone in eastern Arizona, west to Nogales, Prescott, and perhaps Grand Canyon regions (also possibly in Mt. Trumbull area), but distribution between these points not continuous. Also reported as breeding in 1939 at Quitovaquito, western Pima County. Common migrant in eastern Arizona, uncommon west of Baboquivari and Aquarius mountains and the Phoenix region. Winters commonly in weedy farmlands and fairly commonly in grass-brush associations of the Lower Sonoran Zone of central-southern Arizona and the lower Colorado Valley, north and east to the Santa Cruz Valley and the Phoenix area, sparingly to the San Pedro and (casually?) Verde valleys. Has wintered irregularly in grassland at Fort Huachuca (common in late *Feb. 1887*).

404. *Aimophila carpalis* (Coues). Rufous-winged Sparrow. Common resident locally in mixed bunch-grasses and thorn-brush of the Lower Sonoran Zone in central-southern Arizona, from near Oracle and the Tucson region generally west across the Papago Indian Reservation as far as Ventana Ranch and Menager's Dam. Formerly more common and less local, but presently again extending its range, at least to southeastward, where it appeared in 1956-1957 in unexpectedly high places.

405. *Aimophila ruficeps* (Cassin). Rufous-crowned Sparrow; Scott's Sparrow. Common resident of open, grassy and rocky Upper Sonoran Zone hillsides of southern Arizona, north sparingly to parts of the Mogollon Plateau region (summer only?) and west to the Kofa and Ajo Mountains. It is also found sparingly along most of the Grand Canyon, where its range and status are poorly known.

406. *Aimophila botterii* (Sclater). Botteri's Sparrow. Rather uncommon summer resident (no nest yet found within state) from near the southeastern corner of Arizona west to near Sonoita, usually in giant sacaton grass association; also found in Oracle region in *1940*. Formerly much more common, especially before 1895, when it ranged west to the Altar Valley and north to Fort Grant.

407. *Aimophila cassinii* (Woodhouse). Cassin's Sparrow. Common post-breeding summer visitant, and locally and probably irregularly a fairly common winter resident, of the more extensive tall grass areas of the Lower Sonoran Zone in southeastern Arizona. Ranges west to the Coyote and Baboquivari mountains (and exceptionally to west of Growler Mountains in *fall of 1959*), and north to the Gila River; also found at Camp Verde, *July 21, 1916*, and at Monument 180 on the Mexican border in Yuma County, *Aug. 25, 1961*. Has not been found nesting in Arizona or adjacent areas, contrary to A.O.U. Check-list; appears to be absent from early May to late June.

408. *Aimophila quinquestriata* (Sclater and Salvin). Five-striped Sparrow. Known only from a specimen taken at west foot of Santa Rita Mountains, *June 18, 1957*.

409. *"Amphispiza" bilineata* (Cassin). Black-throated Sparrow; Desert Sparrow. Common summer resident of scattered low brush or cactus in arid Sonoran zones throughout the state, scarcer west of the Gila Bend and Organ Pipe Cactus National Monument areas in the south. Winters rather commonly in scattered Lower Sonoran thorn-brush north to the Gila and lower Salt River valleys (casually north of Prescott), thence to the Hoover Dam region, but sparsely west of the Castle Dome Mountains, seldom reaching the extreme lower Colorado River, and absent from brushless grassland in the southeast. Rare transient (and wanderer?) on the Mogollon Plateau.

410. *"Amphispiza" belli* (Cassin). Sage Sparrow. Common to abundant summer resident of open sage-brush on the Navajo Indian Reservation north of the Rio Puerco Valley, west to Echo Cliffs and the Hopi Buttes (and possibly in sage areas farther west?). Summer reports elsewhere probably due to confusion with juvenal *bilineata*. Winters commonly on open ground with sparse brush in southwestern and western Arizona, scarcer farther east and south. Also occurs in winter in dense salt-bush stands. Casual fall transient on the high prairies of the White Mountains.

411. *Junco "aikeni"* Ridgway. "White-winged" Junco. Recorded only in the winter of *1936-7,* when one was taken near Espero in the White Mountains, *Nov. 21,* and at least six were seen (five taken, one banded) near Flagstaff, *Jan. 21 to Feb. 26.* All Juncos are regarded as races of *hyemalis,* the Dark-eyed Junco, except the *phaeonotus* group and *vulcani.*

412. *Junco hyemalis* (Linnaeus). "Slate-colored" Junco; "Cassiar" Junco. Very uncommon but regular winter resident in eastern Arizona, rarer to the west, but ranging all the way to Yuma. Usually single birds with flocks of other "species," chiefly "*oreganus.*"

413. *Junco "oreganus"* (Townsend). "Oregon" Junco; "Pinksided" Junco; "Montana" Junco; "Shufeldt's" Junco; "Thurber's" Junco; "Sierra" Junco. Abundant winter resident of open forests and woods of the Upper Sonoran Zone and above; rather common in the moister, more brushy areas of the Lower Sonoran Zone. Also common in wet years in winter in southwestern desert mountain ranges.

414. *Junco "caniceps"* (Woodhouse). "Gray-headed" Junco; "Red-backed" Junco. Very common summer resident in boreal forests of the Mogollon and Kaibab Plateaus; fairly common summer resident in the adjacent Transition Zone, and on the Coconino Plateau and in the northeast. Reported sparingly south in former years to near Clifton, the Natanes Plateau, the Sierra Ancha, and the Mazatzal and Bradshaw Mountains, but no nest yet found in any of these ranges. Very common winter resident, with the foregoing "species," in eastern Arizona, uncommon farther west to the Colorado River. Common transient across Transition Zone of Mogollon Plateau, rarer in lowlands. Early June specimens were taken at Fort Huachuca, *June 5, 1892,* and in the Guadalupe Mountains, *June 7, 1957.*

415. *Junco phaeonotus* Wagler. Yellow-eyed Junco; Arizona Junco; Mexican Junco. Abundant resident in Transition and boreal zones of southern Arizona, north and west to the Santa Rita, Santa Catalina, Pinal and Graham mountains. Also winters in adjacent Upper Sonoran Zone, rarely reaching the lower edge of Upper Sonoran, as early as *Sept. 18, 1952* (Santa Catalina Mountains). One migrant station outside of breeding range: Whetstone Mountains, *Sept. 26* to Oct. 5, *1907.*

416. *Spizella arborea* (Wilson). Tree Sparrow. Uncommon winter resident in brushy and weedy parts of the Transition Zone, and of Upper Sonoran rivers and farmlands, on and northeast of the Mogollon Plateau, and south and west to Moenave (east of the Grand Canyon), Flagstaff, and upper Black River in the White Mountains; casual to the San Carlos area, Gila Valley, *Jan. 11* and *22, 1937,* and to southern Nevada; once seen inside Grand Canyon. An old report near Tucson unlikely and unsubstantiated.

417. ***Spizella passerina*** (Bechstein). Chipping Sparrow. Abundant summer resident in open parts of the Transition and boreal zones, and rather common in open wooded parts of Upper Sonoran Zone, of northern Arizona, west and south to the Hualapai Mountains, the Prescott region, Payson, and Whiteriver. Also breeds locally in Upper Sonoran woods of the Huachuca and Chiricahua Mountains; supposed June or breeding records farther west not substantiated. Migrates throughout the state; winters abundantly in oak-grasslands (Upper Sonoran Zone) of central-southern Arizona, and commonly in farmlands and the moister parts (grassy or wooded) of Lower Sonoran Zone throughout southern Arizona, north less commonly to Davis Dam (once to Lake Mead, Nevada), the Big Sandy Valley, and the lower Salt and Gila River valleys. May winter casually in Transition Zone of White Mountains (specimens, *late Nov. 1936*); but a specimen from Bridgeport, Verde Valley, *Feb. 24, 1931,* does not justify the statement (A.O.U. Check-list) that it winters at Camp Verde.

418. ***Spizella pallida*** (Swainson). Clay-colored Sparrow. Rare transient, irregularly (or formerly?) more common, from Sonoita Valley east. Irregular west to Tucson, and Altar Valley (Jan. 1956). Specimens have been taken in the upper San Pedro Valley, at Fort Crittenden, near Tucson and Elfrida (in Sulphur Springs Valley), and probably near Bisbee. Found wintering in some numbers just inside Sonora in the San Rafael Valley. Some published records are erroneous.

419. ***Spizella breweri*** Cassin. Brewer's Sparrow. Common summer resident in sage and other tall, dense brush of Upper Sonoran Zone in the northern part of the Navajo Indian Reservation. A colony also in Lower Sonoran Zone, Camp Verde to Fossil Creek, in 1880's. There are single summer records for Flagstaff and Vail (near Tucson), but none for "Fort Whipple" (Prescott), where erroneously stated to breed (A.O.U. Check-list). Common migrant in open parts of the Transition and Sonoran zones statewide. Abundant winter resident in Lower Sonoran Zone except in drier open areas (purer creosote) of the extreme southwest and in uncultivated portions of the Colorado Valley; rarer in its northern edge, but winters to near Kingman and to southern Nevada.

420. ***Spizella atrogularis*** (Cabanis). Black-chinned Sparrow. Fairly common summer resident in Upper Sonoran chaparral across Arizona from east to northwest below the rim of the Mogollon Plateau and west to Hualapai Mountains; south very locally to the Chiricahua and Mule Mountains, and possibly the south end of the Huachuca Mountains (but not in Sonora, as stated in A.O.U. Check-list). Once seen at Flagstaff, Aug. 5, 1947. Fairly common winter resident locally in scattered brush of high Lower Sonoran and low Upper Sonoran hillsides of central-southern Arizona, from the Natanes Plateau and the Santa Catalina Mountains west to the Ajo Mountains, and even in some years to the higher mountains of Yuma County. East of the San Pedro Valley known as a wintering bird only in the Chiricahua Mountains, where locally resident. Casual transient to Colorado River and lower mountains and valleys of Yuma County; unrecorded in open valleys of the southeast, and in the northeast.

421. *Zonotrichia querula* (Nuttall). Harris' Sparrow. Four records, south to Gila Valley: Sacaton, *March 16, 1913;* Moenave, Navajo Indian Reservation, Feb. 19, 1937; Tempe (banded), Nov. 22 to Dec. 23, 1947; and San Carlos, *Nov. 9, 1951.*

422. *Zonotrichia leucophrys* (Forster). White-crowned Sparrow; Gambel's Sparrow; Intermediate Sparrow. Very common transient in brushy places anywhere. Abundant winter resident in weeds about tall brush, principally in farmlands and large washes, throughout the Sonoran zones, though less numerous north of the Mogollon Plateau and east of the Tuba City region. Leaves Mogollon Plateau and Grand Canyon (rims) by early December. Casual (not known to breed) near timberline in the White Mountains, *July 11, 1936;* has been seen casually in summer in Lower Sonoran Zone, where it lingers regularly to *early June.*

423. *Zonotrichia atricapilla* (Gmelin). Golden-crowned Sparrow. Very rare winter visitant in recent years in southern Arizona; specimens from Topock, the Ajo Mountains, near Tucson (two), and an older one from Potholes, California; sight records from Tucson and the Colorado River six miles above Imperial Dam. Two records on migration in northern Arizona, Springerville, Apr. 25, 1953, and near Anita, Coconino County, *Oct. 8, 1956.*

424. *Zonotrichia albicollis* (Gmelin). White-throated Sparrow. Rare winter resident in recent years in Lower Sonoran brush in valleys and foothill canyons (casually in Upper Sonoran Zone) across southern Arizona, except the drier deserts, west and north to the Colorado River (north to Nevada) and the Verde River (at or above its mouth).

425. *Passerella iliaca* (Merrem). Fox Sparrow. Local winter resident, usually in very small numbers except in the Hualapai Mountains (and, in some years, the Kofa and probably adjacent mountains), in dense Sonoran Zone thickets and chaparral in western and southern Arizona, south barely into Sonora. Rare transient along the lower Colorado River and other principal watercourses and canyons, and in the San Francisco and White Mountains regions. Accidental in western Arizona as late as *May 15, 1956* (Kofa Mountains) and as early as Aug. 29, 1938 (below Hoover Dam).

426. *Melospiza lincolnii* (Audubon). Lincoln's Sparrow. Summer resident, perhaps not uncommon, in willows of higher elevations of the White Mountains; possibly also at Mormon Lake. Common migrant in dense low cover bordered by grass or weeds, probably throughout state. Winters rather commonly in dense brush, reeds, and farm hedgerows of the main Lower Sonoran Zone valleys of southern and western Arizona, east and north to near Patagonia, the lower San Pedro Valley, Phoenix, the Davis

Dam area, and casually to Boulder City, Nevada (but not, as stated by A.O.U. Check-list, at Flagstaff nor on San Francisco Mountain).

427. *Melospiza georgiana* (Latham). Swamp Sparrow. Rare and irregular winter visitant in southeastern Arizona west to Tucson, casually west to the Colorado River in southern Arizona (Bill Williams Delta, *Nov. 28, 1952* and *Dec. 23, 1953*). One record from northern Arizona: Tuba City, Dec. 19, 1936. The statement that it occurs only "casually" in Arizona (A.O.U. Check-list) hardly agrees with the facts; from Tucson alone there are four specimens, and twice it has been thought that four individuals were present at the same spot and time; there are additional sight records near and south of Tucson, and additional specimens from Hereford (San Pedro River) and the New Mexico part of the San Simon Cienega.

428. *Melospiza melodia* (Wilson). Song Sparrow. Locally common resident in reed-sedge-brush types along major permanent rivers in southern and western Arizona, east to Safford and to San Bernardino Ranch in extreme southeast; also summer resident along the upper Colorado River (Grand Canyon region) and on permanent brush-lined streams of Upper Sonoran and Transition zones on and adjacent to the Mogollon Plateau (chiefly in the White Mountains region). Rather common winter resident locally at reedy ponds, brushy streams, and farmlands with brushy, weedy edges in the Sonoran Zone valleys and even in low Transition Zone where it does not breed, and probably higher. Very rare transient elsewhere. Greatly reduced as a resident bird in the southeast in the present century because of habitat restriction.

429. *"Rhynchophanes" mccownii* (Lawrence). McCown's Longspur. Rare and irregular (formerly abundant) winter resident in grassy plains and valleys of eastern Arizona; recorded west formerly to the Altar Valley, the Gila Valley south of Phoenix, the Agua Fria River east of Prescott, and the northeast slope of the San Francisco Mountains. Since 1922 recorded only from the Sonoita-Elgin-San Rafael Valley area.

430. *Calcarius lapponicus* (Linnaeus). Lapland Longspur. Apparently very rare, only two acceptable specimens: Meteor Crater, Coconino County, *Nov. 15, 1947;* and Colorado River about seven miles above Imperial Dam, *Nov. 18, 1955.* Also two seen at latter pont, Nov. 11, 1956; and has been seen in winter in western part of Navajo Indian Reservation, where it may be regular. Specimens found in other places (Phoenix, Petrified Forest) were probably brought there by automobiles coming from undetermined regions, perhaps out-of-state, and the latter record should be deleted from A.O.U. Check-list.

431. *Calcarius pictus* (Swainson). Smith's Longspur. One specimen: White Mountains, *Apr. 24, 1953.* No winter record, but possibly not as scarce in Arizona and northern Sonora as this single specimen would imply.

432. *Calcarius ornatus* (Townsend). Chestnut-collared Longspur. Abundant winter resident in the grasslands of southeastern Arizona, and fairly common migrant and winter resident in grasslands farther north, even up to the Canadian Zone (snow-cover permitting). In fall migration occurs in small numbers (singly or in flocks of five or less) more or less regularly, usually on open land near water, nearly statewide west to the Colorado River.

PART 5

RECENT MAMMALS OF ARIZONA

E. Lendell Cockrum
The University of Arizona

The following check list of the species of recent mammals is intended to indicate the diversity and the gross distributional patterns of the mammals of the state. Those who are interested in keys to aid in the identification of these species and in their more exact distributional limits are referred to: E. R. Hall and K. R. Kelson, "The Mammals of North America," Ronald Press, 1958, or to E. L. Cockrum, "The Recent Mammals of Arizona: Their Taxonomy and Distribution," University of Arizona Press, 1960.

Order Marsupialia
Family Didelphidae: Opossums

1. *Didelphis marsupialis* Linnaeus. Opossum. Rare; reported as introduced in southern and central Arizona, although not well established anywhere.

Order Insectivora
Family Soricidae: Shrews

2. *Sorex merriami* Dobson. Merriam's Shrew. Rare; known from high elevations in Coconino and Greenlee counties (7,300-8,700 feet).

3. *Sorex vagrans* Baird. Vagrant Shrew. Common at higher elevations in northern part of the state and from isolated mountains in southern part of the state: Graham, Santa Catalina, Santa Rita, Huachuca and Chiricahua mountains.

4. *Sorex nanus* Merriam. Dwarf Shrew. Rare; known from the Kaibab Plateau and the White Mountains.

5. *Sorex palustris* Richardson. Water Shrew. Rare; known from the White Mountains and Blue Range in east-central Arizona (8,300-9,500 feet).

6. *Notiosorex crawfordi* (Coues). Desert Shrew. Locally common; widely distributed statewide at elevations below 6,000 feet, especially in riparian situations.

Order Chiroptera

Family Phyllostomatidae: Leaf-nosed Bats

7. *Mormoops megalophylla* (Peters). Leaf-chinned Bat. Rare; known from two specimens from Santa Cruz County.

8. *Macrotus californicus* Baird. California Leaf-nosed Bat. Locally common in shallow caves, mine tunnels and under bridges. Occurs widely at lower elevations in the western and southern parts of the state.

9. *Choeronycteris mexicana* Tschudi. Mexican Long-tailed Bat. Uncommon; usually found near the fronts of shallow caves and mine tunnels. Known from Pima, Santa Cruz and Cochise counties.

10. *Leptonycteris nivalis* (Saussure) Long-nosed Bat. Locally common in moist caves. Known from Pinal, Pima, Santa Cruz and Cochise counties.

Family Vespertilionidae: Plain-nosed Bats

11. *Myotis yumanensis* (H. Allen). Yuma Myotis. Locally common. statewide in distribution.

12. *Myotis velifer* (J. A. Allen). Cave Myotis. Locally abundant in summer months at lower elevations (below 5,000 feet) throughout the southern and western parts of the state.

13. *Myotis occultus* Hollister. Arizona Myotis. Rare; known from a few localities below the Mogollon Rim.

14. *Myotis evotis* (H. Allen). Long-eared Myotis. Uncommon in mine tunnels and hollow trees at elevations above 5,000 feet throughout central and northern Arizona.

15. *Myotis keenii* (Merriam). Keen's Myotis. Uncommon at elevations above 5,000 feet in the southeastern part of the state.

16. *Myotis thysanodes* Miller. Fringe-tailed Myotis. Locally common in caves, mine tunnels and buildings at elevations above 5,000 feet throughout the state.

17. *Myotis volans* H. A. Allen. Long-legged Myotis. Uncommon; statewide at elevations above 5,000 feet.

18. *Myotis californicus* Audubon and Bachman. California Myotis. Locally common throughout the state.

19. *Myotis subulatus* (Say). Small-footed Myotis. Uncommon but distributed throughout the state.

20. *Lasionycteris noctivagans* (Le Conte). Silver-haired Bat. Uncommon solitary tree dwelling bat found throughout the state at elevations above 5,000 feet.

21. *Pipistrellus hesperus* (H. Allen). Western Pipistrelle. Common throughout the state.

22. *Eptesicus fuscus* (Palisot de Beauvois). Big Brown Bat. Locally common throughout the state.

23. *Lasiurus borealis* (Muller). Red Bat. Uncommon solitary tree bat throughout the state in the region of trees.

24. *Lasiurus cinereus* (Palisot de Beauvois). Hoary Bat. Uncommon tree dwelling bat found throughout the state in the region of trees.

25. *Dasypterus ega* (Gervais). Yellow Bat. Rare; known only from two specimens from Tucson.

26. *Plecotus townsendii* (Cooper). Lump-nosed Bat. Locally common throughout the state at elevations above 5,000 feet; rare at lower elevations.

27. *Plecotus phyllotis* (G. M. Allen) Mexican Big-eared Bat. Rare; known from Graham, Cochise and Coconino counties, from the oak zone.

28. *Euderma maculata* (J. A. Allen). Spotted Bat. Extremely rare; known from four specimens, Maricopa and Yuma counties.

29. *Antrozous pallidus* Le Conte. Pallid Bat. Locally common throughout the state.

Family Molossidae: Free-tailed Bats

30. *Tadarida brasiliensis* (I. Geof. St.-Hilaire). Mexican Free-tailed Bat. Locally abundant throughout the state, especially at elevations below 5,000 feet.

31. *Tadarida femorosaca* (Merriam). Pocketed Free-tailed Bat. Rare; found at lower elevations in the western and southern part of the state.

32. *Tadarida molossa* (Pallas). Big Free-tailed Bat. Rare; statewide, mainly at elevations below 5,000 feet.

33. *Eumops perotis* (Schinz). Western Mastiff Bat. Rare; in small colonies in rock crevices at lower elevations in the western and southern part of the state.

34. *Eumops underwoodi* Goodwin. Underwood's Mastiff Bat. Rare; known only from the region of Sasabe, Pima County.

Order Lagomorpha

Family Leporidae: Hares and Rabbits

35. *Lepus alleni* (Mearns). Antelope Jack Rabbit. Occurs in the central third of the southern half of the state.

36. *Lepus californicus* Gray. Black-tailed Jack Rabbit. Statewide.

37. *Sylvilagus floridanus* (J. A. Allen). Eastern Cottontail. Wooded regions in the Hualapai mountains, the region south of the Mogollon Rim and in scattered mountain ranges to the southward.

38. ***Sylvilagus nuttallii*** (Bachman). Nuttall's Cottontail. At higher elevations of northeastern and northwestern part of the state (6,500-9,500 feet).

39. ***Sylvilagus audubonii*** (Baird). Desert Cottontail. Common at elevations below 6,000 feet throughout the state.

Order Rodentia

Family Sciuridae: Squirrels and Allies

40. ***Cynomys ludovicianus*** (Ord). Black-tailed Prairie Dog. Now probably extinct; formerly locally common in grasslands of the southeastern part of the state.

41. ***Cynomys gunnisoni*** (Baird). Gunnison's Prairie Dog. Formerly widely distributed in grasslands of central and northeastern Arizona; now greatly reduced in numbers and range.

42. ***Citellus tridecemlineatus*** (Mitchell). Thirteen-lined Ground Squirrel. Known only from grassy areas in the White Mountains.

43. ***Citellus spilosoma*** (Bennett). Spotted Ground Squirrel. Locally common in grasslands in eastern, central and northern parts of the state.

44. ***Citellus variegatus*** (Erxleben). Rock Squirrel. Statewide, especially at elevations below 6,000 feet.

45. ***Citellus harrisii*** (Audubon and Bachman). Harris Antelope Squirrel. Southern and western parts of the state at elevations below 6,000 feet.

46. ***Citellus leucurus*** (Merriam). White-tailed Antelope Squirrel. Northern parts of the state at elevations below 6,000 feet.

47. ***Citellus tereticaudus*** (Baird). Round-tailed Ground Squirrel. Lower Sonoran Life-zone of the southwestern part of the state (below 3,200 feet).

48. ***Citellus lateralis*** (Say). Golden-mantled Ground Squirrel. Higher elevations (above 6,500 feet) of northern Arizona.

49. ***Eutamias minimus*** (Bachman). Least Chipmunk. Common at higher elevations in Chuska and White Mountains (8,000-11,000 feet).

50. ***Eutamias quadrivittatus*** (Say). Colorado Chipmunk. Northeastern Arizona (6,000-8,000 feet).

51. ***Eutamias cinereicollis*** (J. A. Allen). Gray-collared Chipmunk. Common in higher mountain country above Mogollon Rim (7,500-11,300 feet).

52. ***Eutamias umbrinus*** (J. A. Allen). Uinta Chipmunk. Restricted to Kaibab Plateau (7,000-8,500 feet).

53. *Eutamias dorsalis* (Baird). Cliff Chipmunk. Widely distributed in the mountains of northern, central and southeastern Arizona (4,700-9,400 feet).

54. *Sciurus aberti* Woodhouse. Abert's Squirrel. Restricted to Ponderosa pine areas north of the Mogollon Rim. Introduced into the Graham, Hualapai, Horsethief, Pinal, Bradshaw and Santa Catalina mountains, 5,000-9,000 feet. The Kaibab squirrel is here considered to be a subspecies of *S. aberti.*

55. *Sciurus arizonensis* (Coues). Arizona Gray Squirrel. Mountainous areas along and north of the Mogollon Rim and the Santa Catalina, Rincon and Huachuca mountains on the southeastern part of the state (4,800-8,000 feet).

56. *Sciurus apache* J. A. Allen. Apache Squirrel. Known only from the Chiricahua mountains (5,200-7,500 feet).

57. *Tamiasciurus hudsonicus* (Erxleben). Red Squirrel. Widely distributed in the mountains of northern and central Arizona (6,750-11,000 feet).

Family Geomyidae: Pocket Gophers

58. *Thomomys bottae* (Eydoux and Gervais). Valley Pocket Gopher. Widely distributed throughout the state at all elevations.

59. *Thomomys umbrinus* (Richardson). Southern Pocket Gopher. Known from the Santa Rita, Patagonia, Huachuca and Pajarito mountains (4,300-6,200 feet).

60. *Thomomys talpoides* (Richardson). Northern Pocket Gopher. Known only from the Kaibab plateau and the Lukachukai mountains (7,500-9,000 feet).

Family Heteromyidae: Kangaroo Rats and Pocket Mice

61. *Perognathus flavus* Baird. Silky Pocket Mouse. Locally common in grasslands throughout the state (2,900-6,500 feet).

62. *Perognathus apache* Merriam. Apache Pocket Mouse. Known from grassland areas of the northeastern part of the state; Coconino, Navajo and Apache counties (4,500-6,800 feet).

63. *Perognathus longimembris* (Coues). Little Pocket Mouse. Known from scattered localities in the western part of the state: Yuma, Pinal, Maricopa, Coconino and Mohave counties (500-4,500 feet).

64. *Perognathus amplus* Osgood. Arizona Pocket Mouse. Locally common in desert areas on south-central, western and north-central parts of the state (500-5,100 feet).

65. *Perognathus parvus* (Peale). Great Basin Pocket Mouse. Known from northern Arizona north of the Colorado River (4,900-6,250 feet).

66. *Perognathus formosus* Merriam. Long-tailed Pocket Mouse. Known from northern Arizona north of the Colorado River (1,500-5,400 feet).

67. *Perognathus baileyi* Merriam. Bailey's Pocket Mouse. Widely distributed in the southern part of the state (900-4,700 feet).

68. *Perognathus hispidus* Baird. Hispid Pocket Mouse. Locally common in grasslands of southeastern part of the state; an isolated population tion occurs near Camp Verde (3,200-5,000 feet).

69. *Perognathus penicillatus* Woodhouse. Desert Pocket Mouse. Widely distributed in desert and low grasslands of southern and western Arizona (120-5,200 feet).

70. *Perognathus intermedius* Merriam. Rock Pocket Mouse. Widely distributed in rocky areas in the Colorado River valley, western and southern Arizona (120-6,000 feet).

71. *Dipodomys spectabilis* Merriam. Banner-tailed Kangaroo Rat. Locally common in grasslands of southeastern Arizona (1,300-5,000 feet).

72. *Dipodomys merriami* Mearns. Merriam's Kangaroo Rat. Widely distributed in western and southern parts of the state (120-5,000 feet).

73. *Dipodomys ordii* Woodhouse. Ord's Kangaroo Rat. Widely distributed in grasslands of northern and eastern parts of the state (2,700-7,000 feet).

74. *Dipodomys microps* (Merriam). Chisel-toothed Kangaroo Rat. Restricted to lower elevations north of the Colorado River (3,500-5,400 feet).

75. *Dipodomys deserti* Stephens. Desert Kangaroo Rat. Restricted to sandy areas of the western part of the state (120-1,800 feet).

Family Castoridae: Beaver

76. *Castor canadensis* Kuhle. Beaver. Formerly widespread in all of the permanent streams of the state; now restricted in distribution.

Family Cricetidae: Native Rats and Mice

77. *Onychomys leucogaster* (Wied-Neuwied). Northern Grasshopper Mouse. Widely distributed in the grasslands of the northern and eastern parts of the state (3,100-7,000 feet).

78. *Onychomys torridus* (Coues). Southern Grasshopper Mouse. Widely distributed in the western and southern parts of the state (120-5,000 feet).

79. *Reithrodontomys montanus* (Baird). Plains Harvest Mouse. Rare; known only from a few specimens from the grasslands of Pima and Cochise counties.

80. *Reithrodontomys megalotis* (Baird). Western Harvest Mouse. Statewide (120-8,000 feet).

81. *Reithrodontomys fulvescens* J. A. Allen. Fulvous Harvest Mouse. Known only from eastern Pima, western Cochise and Santa Cruz counties (2,700-5,300 feet).

82. *Baiomys taylori* (Thomas). Northern Pygmy Mouse. Rare in the southeastern grasslands of the state.

83. *Peromyscus crinitus* (Merriam). Canyon Mouse. Widely distributed in northern and western parts of the state, usually restricted to rocky habitats (1,300-6,800 feet).

84. *Peromyscus eremicus* (Baird). Cactus Mouse. Widely distributed in western and southern Arizona (120-6,000 feet).

85. *Peromyscus merriami* Mearns. Merriam's Mouse. Known from scattered localities in Pinal, Pima and Santa Cruz counties (1,600-3,600 feet).

86. *Peromyscus maniculatus* (Wagner). Deer Mouse. Statewide (120-11,400 feet).

87. *Peromyscus leucopus* (Rafinesque). White-footed Mouse. Known from eastern and central parts of the state (2,300-6,500 feet).

88. *Peromyscus boylii* (Baird). Brush Mouse. Known from higher elevations throughout the state (3,000-9,000 feet).

89. *Peromyscus pectoralis* Osgood. White-ankled Mouse. Known from only a few immature specimens from the Huachuca Mountains. These may actually be *P. boylii.*

90. *Peromyscus truei* (Shufeldt). Pinyon Mouse. Widely distributed in the northern part of the state and the Chiricahua Mountains (4,500-7,000 feet).

91. *Peromyscus nasutus* (J. A. Allen). Rock Mouse. Known only from scattered localities in northeastern Arizona and from the Chiricahua Mountains (4,500-7,500 feet).

92. *Sigmodon hispidus* Say and Ord. Hispid Cotton Rat. Known from scattered riparian and grassland areas in southern part of the state (120-5,000 feet).

93. *Sigmodon minimus* Mearns. Least Cotton Rat. Known from scattered grassland areas in southeastern Arizona (3,800-5,200 feet).

94. *Sigmodon ochrognathus* V. Bailey. Yellow-nosed Cotton Rat. Known from scattered localities in Cochise and Santa Cruz counties (6,100-8,400 feet).

95. *Neotoma albigula* Hartley. White-throated Wood Rat. Widely distributed at elevations below 7,000 feet throughout all of the state south of the Colorado River (120-8,000 feet).

96. *Neotoma cinerea* (Ord). Bushy-tailed Wood Rat. Known from the northwestern part of the state and from the Kaibab Plateau (6,000-7,000 feet).

97. *Neotoma mexicana* Baird. Mexican Wood Rat. Known from higher elevations throughout the state south of the Colorado River (5,000-9,200 feet).

98. *Neotoma lepida* Thomas. Desert Wood Rat. Known from north of the Colorado River and from lower elevations in the western part of the state (1,100-6,000 feet).

99. *Neotoma stephensi* Goldman. Stephens' Wood Rat. Widely distributed in northern and central Arizona south of the Colorado River (4,800-7,000 feet).

100. *Clethrionomys gapperi* (Vigors). Gapper's Red-backed Mouse. Known from higher elevations of the White Mountains and Blue Range (8,300-9,600 feet).

101. *Microtus montanus* (Peale). Montane Vole. Known from higher elevations of the White Mountains and the Blue Range (6,900-9,500 feet).

102. *Microtus longicaudus* Merriam. Long-tailed Vole. Known from higher elevations of the Kaibab Plateau, San Francisco, Graham, White and Lukachukai mountains and the Blue Range (6,700-11,000 feet).

103. *Microtus mexicanus* (Saussure). Mexican Vole. Widely distributed at higher elevations above the Mogollon Rim and in the Navajo and Hualapai mountains (6,500-11,500 feet).

104. *Ondatra zibethicus* (Linnaeus). Muskrat. In the lower Colorado and upper parts of the Gila river systems.

Family Muridae: Introduced Rats and Mice

105. *Rattus norvegicus* (Berkenhout). Norway Rat. Introduced but not common; found only around larger towns and cities.

106. *Rattus rattus* (Linnaeus). Black Rat. Introduced but not common; may not be established in the state at present.

107. *Mus musculus* Linnaeus. House Mouse. Introduced; often around dwellings and occasionally occurring as feral populations.

Family Zapodidae: Jumping Mice

108. ***Zapus princeps*** J. A. Allen. Western Jumping Mouse. Restricted to the higher portions of the White Mountains (770-8,600 feet).

Family Erethizontidae: Porcupines

109. ***Erethizon dorsatum*** Linnaeus. Porcupine. Probably statewide but more common in wooded areas (3,000-9,000 feet).

Order Carnivora
Family Canidae: Dogs and Allies

110. ***Canis latrans*** Say. Coyote. Statewide (120-9,100 feet).

111. ***Canis lupus*** Frisch. Gray Wolf. Formerly throughout the eastern portions of the state, at present rare or approximately extinct.

112. ***Vulpes macrotis*** Merriam. Kit Fox. Widely distributed at lower elevations throughout the southern part of the state (120-5,000 feet).

113. ***Urocyon cinereoargenteus*** (Schreber). Gray Fox. Statewide with the possible exception of the northeastern portion (120-5,800 feet).

Family Ursidae: Bears

114. ***Euarctos americanus*** (Pallas). Black Bear. Formerly common throughout the mountainous areas of the state, now greatly reduced in numbers and distribution.

115. ***Ursus horribilis*** Ord. Grizzly Bear. Formerly throughout the mountainous areas of the state, now extinct in Arizona.

Family Procyonidae: Raccoons and Allies

116. ***Bassariscus astutus*** (Lichenstein). Ringtail. Statewide (120-6,500 feet).

117. ***Procyon lotor*** (Linnaeus). Raccoon. Riparian situations along the Colorado, Little Colorado and Gila river systems and in the grasslands of the southeastern portion of the state (120-6,900 feet).

118. ***Nasua narica*** (Linnaeus). Coati. In woodland situations in the Graham, Chiricahua, Huachuca, Patagonia and Pena Blanca mountains (5,000-7,500 feet).

Family Mustelidae: Weasels, Skunks and Allies

119. ***Mustela frenata*** Lichtenstein. Long-tailed Weasel. Higher elevations of central and eastern Arizona (4,000-11,200 feet, probably extinct at lower elevations today).

120. ***Mustela nigripes*** (Audubon and Bachman). Black-footed Ferret. Formerly associated with prairie dog towns of northeastern Arizona, no recent records.

121. ***Taxidea taxus*** (Schreber). Badger. Statewide (120-7,000 feet).

122. ***Spilogale putorius*** (Linnaeus). Spotted Skunk. Probably statewide (120-7,000 feet).

123. ***Mephitis mephitis*** (Schreber). Striped Skunk. Statewide (300-9,000 feet).

124. ***Mephitis macroura*** (Lichtenstein). Hooded Skunk. Southeastern part of the state (2,000-6,000 feet).

125. ***Conepatus mesoleucus*** Lichtenstein. Hog-nosed Skunk. Southeastern part of the state (2,000-6,000 feet).

126. ***Lutra canadensis*** (Schreber). River Otter. Formerly in all of the larger permanent river systems; now rare.

Family Felidae: Cats

127. ***Felis onca*** Linnaeus. Jaguar. Probably formerly rare throughout the state. Today an occasional individual is found in the southern part of the state.

128. ***Felis pardalis*** Linnaeus. Ocelot. Formerly southeastern Arizona as far north as Fort Verde; no recent records.

129. ***Felis concolor*** Linnaeus. Mountain Lion. Statewide (200-8,000 feet).

130. ***Felis yagouaroundi*** Fischer. Jaguarundi. Rare in the southern part of the state; no recent records.

131. ***Lynx rufus*** (Schreber). Bobcat. Statewide (120-9,300 feet).

Order Artiodactyla

Family Tayassuidae: Javelinas

132. ***Pecari tajacu*** (Linnaeus). Javelina. Southeastern and central parts of the state (1,200-6,000 feet).

Family Cervidae: Deer and Allies

133. ***Cervus canadensis*** (Erxleben). Elk. Formerly probably occurred in most of the higher mountains of the state; was exterminated and reintroduced (in 1913); presently occurs at higher elevations in the central part of the state.

134. ***Odocoileus hemionus*** (Rafinesque). Black-tailed or Mule Deer. Statewide, but not of uniform distribution (250-9,000 feet).

135. ***Odocoileus virginianus*** (Zimmermann). White-tailed Deer. Southeastern Arizona (1,200-9,000 feet).

Family Antilocapridae: Prong-horned Antelope

136. ***Antilocapra americana*** (Ord). Prong-horned Antelope. Formerly widely distributed in grassland areas throughout the state; presently restricted to areas of favorable habitat.

Family Bovidae: Cows, Sheep and Allies

137. ***Ovis canadensis*** Shaw. Bighorn. Probably formerly statewide in mountainous or rocky situations; presently restricted to scattered low desert mountains.

INDEX

LANDSCAPES AND HABITATS

COMMON NAMES

Fishes, pages 133-151

Amphibians and Reptiles, pages 153-174

Birds, pages 175-248

Mammals, pages 249-259

FAMILIES AND GENERA